Cities and Race

D0147808

Today, in the shadows of gleaming downtown skyscrapers and showy gentrified neighborhoods, conditions in many impoverished black ghettos in America's rust belt have substantially worsened. Leaders and residents in these communities struggle to acquire the resources to upgrade their communities, but contest a formidable obstacle: the accelerated push to make and protect downtown revitalized landscapes of consumption, pleasure, and affluent residency.

Cities and Race comprehensively explores this new black ghetto reality and discusses and explains:

- The rise of a new kind of black ghetto termed "the glocal ghetto"
- The reality of a new third wave of black ghetto marginalizing since 1945 in public policy and popular discourse in America
- A new political-economic force that triggers the production of this new ghetto, "the global trope"
- The ascendant characteristics of this new ghetto: a deepened poverty of its residents, a new denigrating pattern of representation assigned to residents and these communities, and a continued connection of this space to the prison-industrial complex in America
- The bolstered role that local politics plays in producing these new ghetto spaces.

Cities and Race concludes, in rich and original detail, that America has now spawned a new kind of ghetto that has become more impoverished and more inpugned as the now crystallized zone of human discard in "the global era."

David Wilson is Professor of Geography at the University of Illinois at Urbana-Champaign.

Questioning Cities

Edited by Gary Bridge, *University of Bristol*, UK and Sophie Watson, *The Open University*, UK

The "Questioning Cities" series brings together an unusual mix of urban scholars under the title. Rather than taking a broadly economic approach, planning approach or more socio-cultural approach, it aims to include titles from a multi-disciplinary field of those interested in critical urban analysis. The series thus includes authors who draw on contemporary social, urban and critical theory to explore different aspects of the city. It is not therefore a series made up of books which are largely case studies of different cities and predominantly descriptive. It seeks instead to extend current debates, through in most cases, excellent empirical work, and to develop sophisticated understandings of the city from a number of disciplines including geography, sociology, politics, planning, cultural studies, philosophy and literature. The series also aims to be thoroughly international where possible, to be innovative, to surprise, and to challenge received wisdom in urban studies. Overall it will encourage a multi-disciplinary and international dialogue always bearing in mind that simple description or empirical observation which is not located within a broader theoretical framework would not – for this series at least – be enough.

Published:

Global Metropolitan
John Rennie Short

Reason in the City of Difference
Gary Bridge

In the Nature of Cities: Urban political ecology and the politics of urban metabolism
Erik Swyngedouw, Maria Kaika, Nik Heynen

Ordinary Cities: Between modernity and development
Jennifer Robinson

Urban Space and Cityscapes
Christoph Lindner

City Publics: The (dis)enchantments of urban encounters
Sophie Watson

Small Cities: Urban experience beyond the metropolis
David Bell and Mark Jayne

Cities and Race: America's new black ghetto
David Wilson

Cities in Globalization: Practices, policies and theories
Peter J Taylor, Ben Derudder, Piet Saey and Frank Witlox

Cities and Race

America's new black ghetto

David Wilson

LONDON AND NEW YORK

First published 2007
by Routledge
2 Park Sq, Milton Park, Abingdon, Oxon OX14 4RN

Simultaneously published in the USA and Canada
by Routledge
270 Madison Ave, New York, NY 10016

Routledge is an imprint of the Taylor & Francis Group, an informa business

© 2007 David Wilson

Typeset in Times New Roman by Graphicraft Limited, Hong Kong
Printed and bound in Great Britain by Antony Rowe Ltd, Chippenham, Wiltshire

All rights reserved. No part of this book may be reprinted or reproduced or utilised in any form or by any electronic, mechanical, or other means, now known or hereafter invented, including photocopying and recording, or in any information storage or retrieval system, without permission in writing from the publishers.

British Library Cataloguing in Publication Data
A catalogue record for this book is available from the British Library

Library of Congress Cataloging in Publication Data
Wilson, David, 1956–
Cities and race : America's new black ghetto / David Wilson
p. cm. – (Questioning cities series)
Simultaneously published in the USA and Canada by Routledge.
Includes bibliographical references and index.
ISBN 0-415-35805-1 (hardcover : alk. paper) – ISBN 0-415-35806-X (softcover : alk. paper) 1. Urban poor–United States. 2. African Americans–Economic conditions. 3. African Americans–Social conditions. 4. Inner cities–United States. 5. Urban policy–United States. 6. United States–Race relations. I. Title. II. Series.
HV4045.W54 2006
307.3′3660973–dc22
2006009658

ISBN 10: 0-415-35805-1 (hbk)
ISBN 10: 0-415-35806-X (pbk)
ISBN 10: 0-203-00410-4 (ebk)

ISBN 13: 978-0-415-35805-7 (hbk)
ISBN 13: 978-0-415-35806-4 (pbk)
ISBN 13: 978-0-203-00410-4 (ebk)

Contents

Tables

Preface

In prevailing economic and social times, the specter of poverty and squalor deepens in U.S. cities. This book is about the contemporary plight of America's most impoverished urban communities: black ghettos in the rust belt. These communities in this region continue to be profoundly battered by waves of conservative (neoliberal) programs and policies amid a supposed resuscitation of many of these cities in recent years. As narrow segments of the population in these cities have gained (i.e. the wealthy and parts of the middle class), the plight of tens of thousands of low- and moderate-income people has, in particular, worsened. Today, unemployment, underemployment, hopelessness, and poverty sear the inner cities of Chicago, Cleveland, Detroit, Philadelphia, St. Louis, and the like. Yet, while local media shine the spotlight on new glistening downtowns and rows of gentrified neighborhoods, the plight of these poorer communities are brutally cast into darkness via a prominent treatment: by being derogated in brief but colorful and caricatured rhetorical flourish.

This book's concern is with two things: the recent processes that have spearheaded this decline in these ghettos and their current realities. On both counts, this work builds on recent studies that finger the impacts of a growing "neoliberal" conservatism and government at all levels (Anyon 1997; Brenner and Theodore 2002). This work I build on identifies the dilemmas of diverse government policies and programs, supported to different degrees by Americans, from Workfare to housing vouchers to cuts in block grant spending. I believe that this work is important – the post-1980 neoliberal ascendancy has undeniably wrought damage on these communities. But I suggest that it is still incomplete and partial: first, there is a crucial context that propels this neoliberal drive, what I call the global trope, which has eluded the attention of too many observers. And second, a new ghetto form has emerged from this, what I will call the glocal ghetto, that is now represented differently in public discourse, viewed differently by the public, treated differently by policy, and exhibits an unparalleled and new kind of marginalization. Let me explain.

As I suggest in this book, a key rhetorical trigger since 1990 drives this decline: the cultivated fear and obsession with the dawn of a supposed new era – globalization ("the global trope") – and its impact has been nothing less than to create a new distinctive ghetto entity. This intense rhetoric, now forcefully and colorfully elaborated in local settings, is a strategic discourse that helps mobilize and put into play such destructive forces as "trickle-down" redevelopment, retrenchment of social service provision, and a stepped-up incendiary rhetoric about the "unproductive" poor. This global-speak follows a logic: it helps powerful growth machines (i.e. assemblages of prominent builders, developers, local government, Realtors, and the media who push for a unified vision of city growth) direct a desired physical and social restructuring. As I discuss, in the post-1990 era of "hyper real-estate capital accumulation," this strategy is not surprising. Black ghettos, key functional units in this, become reinforced as something pernicious to its residents: warehousing zones for "contaminants" of local property values.

That these ghettos currently stash and segregate the people and land-uses deemed contaminants to real-estate markets is nothing new. These spaces performed this purpose prior to 1990, and indeed have a long history steeped in this. However, this trend has recently accelerated, fueled by a new excoriating, decisive rhetoric that has gained a new legitimacy. The catalyzing force in this, the global trope, now provides a widely accepted justification to normalize this isolation of these neighborhoods and offer these destructive programs and policies. While this rhetoric ultimately legitimates a potentially controversial restructuring, these programs and polices further marginalize and isolate these ghettos as stepped-up, targeted zones of human discard. Given this new powerful rhetoric and its effects, I refer here to this evolved, new ghetto as a distinctly new kind of space, "the glocal ghetto."

Yet there is a paradox to all of this. Despite a dearth of media attention, these areas continue to haunt Rust Belt America. These spaces and populations are now, more than ever, imagined as the frightening and resistant ill in cities that will not go away. They continue to be widely seen in melodramatic terms: as culturally downtrodden places replete with ominous characters (roaming gangs, distraught welfare mothers, hustling pimps and prostitutes) and extreme states of being (all-dangerous streets, imploding families, terrifying parks). So marked out and understood, the process of ghetto intensification becomes sanitized: the systematic isolation of these spaces and populations grows, which, paraphrasing Judith Walkowitz (1993), is seen to vindicate virtue over vice. In this setting, I suggest, it is imperative to understand more deeply these spaces and populations – what they are really like, what produces them, and who is doing the sculpting.

How does this study fit into the evolving realm of research on U.S. cities? I believe that it is ultimately in line with but also at odds with a new academic focus: documenting the global's impacts on cities. On the one hand, the study is conventional and suggests that the power and clout of an analytic object – the global – has reached to the furthest corners of these cities. As suggested, every inch of these ghettos, the parks, houses, streets, institutions, and people, bear this influence. This finding thus conforms to most studies about globalization that emphasize the profound reach and restructuring of this omnipresent force. In this standard template, paraphrasing Brenner, Jessop et al. (2003), places and regions cannot escape the global's tenacious grasp. All that exists gets irreparably sculpted and modified by this kind of penetrating cultural and economic "telos" on the move.

But, on the other hand, the study departs from this standard template: it asserts that this global clout has been especially influential as a rhetoric. Public proclamations of new global times transmit a disciplining code of a new ominous world for everyone to assimilate. At its core is the growth machine desire to command and steer physical and social change – in city after city – in evolving capitalist times. City growth machines find themselves in a new economic circumstance (hyper real-estate accumulation), a new political reality (neoliberal era), and a new social setting (vague public notions of an ominous global world). Drawing out this vague global conception – grounding it, empiricizing it, displaying its supposed effects – works through a kind of "opportunity structure" that is logical. Tweaking an old adage, if globalization did not exist, humans would have to create it, for much political mileage can be gained from this as a decisive offering. In the end, it is globalization as an image and a fear, as much as a reality, which today activates people and institutions in rust belt cities to take actions that deepen deprivation and decline in these black ghettos.

The chapters in the book proceed chronologically. Part I, consisting of three chapters, discusses the basics of this study and the emergence of the glocal black ghetto. Chapter one covers the theoretical tools and techniques used in this study to dissect these ghettos. Its emphasis is on explicating the theoretical perspective used: racial economy. Chapter two examines the rise and formation of this ghetto across the twentieth century. From small enclaves rooted in early industrialization and scattered inter-regional migration streams, it is chronicled, an economic and political logic propelled a rapid growth. Chapter three documents the intricacies of the global trope as a complex rhetoric at the center of this space's post-1990 intensification. This global rhetoric is examined as a nuanced kind of text that belies its common presentation as simple and "plain-speaking."

Part II examines the current dynamics of the glocal black ghetto. Chapter four chronicles the ghetto population's current characteristics socially, economically, and institutionally. Chapter five focuses on the ghettos' sustaining by recent policy initiatives under George W. Bush. Both chapters, in aggregate, provide a sense of what these spaces and population are currently like and the most recent set of forces that act upon them. Part IV of the book, the final section, examines this ghetto as an active and evolving place. Chapter six chronicles ongoing efforts by residents and institutions to stem the tide of decline and neglect, focusing on diverse types of community resistance and confrontation. Chapter seven follows from this and discusses the heart of the current crisis in these communities and what needs to be done as ameliorative strategy.

I would like to thank David Adcock, Ian Cook, Colin Flint, Vinny Pattison, Jan Pieterse, Kevin Ward, and the World University Network Human Geography Seminar participants for reading and posing important questions about the text. Andrew Mould and Zoe Kruze of Routledge have been wonderfully supportive of this project from start to completion. Finally, this book is dedicated to two very special people, "Dr. Dunkenstein" (Andrew Wilson) and "the Angelic Shootist" (Matt Wilson), whose lives on and off the basketball court never fail to revive my spirit.

I.

Glocal black ghetto emergence

1 Introduction

THE FRAME

Today, in the shadows of gleaming downtown skyscrapers and showy gentrified neighborhoods, many impoverished black ghettos in America's Rust Belt have substantially worsened (Wacquant 2002, 2002a).[1] These ghettos, frequently found within five to ten minutes drive of investment-energized downtowns, might as well be in another universe.[2] Leaders and residents struggle to acquire the resources to upgrade their communities, but face a formidable obstacle: the accelerated push to make and protect downtown revitalized landscapes of consumption, pleasure, and affluent residency. New redevelopment zones (e.g. the Loop-Gentrification Complex (Chicago), the Circle Centre Mall Axis (Indianapolis), Soulard-Gentry Boulevard (St. Louis), and the Public Square-Historic Gateway Cluster (Cleveland)), have emerged as hyped revitalization icons for what their cities ostensibly can and need to become. In this context, black ghettos, from the gaze of many planners and growth-advocates, simply do not rate.

The thesis of this book clarifies this new black ghetto reality: that these areas more deeply bleed with a bolstered functional logic ascribed to them, to warehouse "contaminants" in the "global-compelled" city restructuring. While these ghettos in Cleveland, Detroit, St. Louis, Chicago, and the like have always warehoused the racial poor and been seared by negative representations, these aspects have accelerated since 1990. As this book documents, deepened neoliberal physical and social restructuring in these cities has created a startlingly new black ghetto entity.[3] Now, a more pronounced material and symbolic deprivation marks these areas under a post-war "third-wave" of black ghetto marginalization. These residents, in expedient processes, are both materially battered and symbolized – understood around a new debilitating theme of hopelessly pathological and destructively "consumptive." Black ghettos, once again but in a new way, are built into the ground, embedded in social relations, and plugged into circuitries of economy and politics.

But what is the third wave of ghetto marginalization that is central to this exploration? This wave, a post-1990 phenomenon, socially and spatially isolates these spaces (via discourses and practices) to make profitable "global-competitive" economic spaces for real-estate capital (a post-war privileged coalition of prominent builders, developers, and Realtors in city policy that has always been entangled with local elite dreams for profit, prestige, and civic improvement). The previous wave, the second, was an early and mid 1980s activating of Reagan's "welfare-ghetto" rhetoric by local growth machines (striking out to assist real-estate capital) to fortify and expand the newest accumulation apparatus: frontier gentrification (Wilson 2005). Yet both have roots in a 1950s and 1960s first wave of black ghetto marginalization whose central analytical object, "the negro slum," purportedly needed isolating or eradicating to economically galvanize cities (Tabb 1974). Whereas the second wave pivoted around nurturing incipient revitalization spaces, the first wave centered upon the use of the urban renewal bulldozer to boldly re-make downtowns. In each case, these black ghettos have felt the wrath of something powerful: punitive, perpetually faltering city economies.

It follows that these ghettos today, despite other assertions, are anything but absent from capital's thoughts and mainstream discourse. In a widespread myth, the ascendant neoliberal 1980s (fueled by Reagan's "Welfare Queen" oratory) powerfully marginalized these spaces and populations, and now erases them from the public mind. In common discourse and daily thought, it is said, they are now forgotten and left to rot.

This book paints a different portrait: that these populations and spaces are still painstakingly managed, particularly by growth machines (amalgams of builders, developers, Realtors, the local state, and the media that push a unified vision of city growth) and the police apparatus. While national rhetoric has lessened this demonizing, widely substituting "commonsense" neoliberal oratory for raw portrayals of atavistic people and spaces, local rhetoric seamlessly deepens this. The sources of this demonizing today, thus begin less with oratory from familiar voices – presidents, think-tank hotheads, and incendiary national columnists – than with local politicians, planning reports, mayoral utterances, and real-estate moguls.

In elaborating this thesis, the book chronicles a crucial catalyst to this third-wave of ghetto ravaging: the recent fear of and obsession with a supposed new era – globalization. This elaborate rhetoric, served up heavily now in local settings, has been a key trigger to mobilize and put into play crucial ghetto-destroying forces (targeting of government resources to cultivate a robust entrepreneurial city, retrenching the local welfare apparatus, rhetorically attacking these populations and spaces). This rhetoric, which I call "the global trope," is framed by and extends neoliberal principles and designs (especially the notions of the private-market as determinant of social and land-use outcomes and the retrenchment of social welfare) to systematically re-make these cities. The global trope, in this frame, is served up as a frank and blunt package of truths about city realities and needs that can no longer be suppressed. In assertion, its pleas correspond to core truths; deft interpreters read and respond to clear truths as a policy prescriptive, progressive human intervention onto a turbulent and fragile city.

The rhetoric of the global trope has thus been a perceptual apparatus with profound material effects. It has served up a digestible reality that, following Robin Wagner-Pacifici (1994), guides construction of programs and policies by making certain actions thinkable and rational and others not. Imposed webs of meanings, like symbolic cages, build bars around senses of reality that place gazes within discrete and confining visions. One reality is ultimately advanced while alternatives are purged. Here is Mikhael Bakhtin's (1981) implicit dialogue with other points of view, the simultaneity of asserting one vision and annihilating others. This strategic affirmation and rebuke, forwarding what exists and what does not, continues to make this rhetorical formation a fundamental instrument of power. As this apparatus has resisted and beaten back competitive visions of city and societal realities, even as it is contested and struggled against, it grows stronger in numerous rust belt cities.

At this rhetoric's core, a supposed new hyper-competitive reality makes rust belt cities easily discardable as places of investment, production, and business. These once enclosed and confident containers of the economic, in the rhetoric, have recently become porous and leaky landscapes rife with a potential for dramatic economic hemorrhaging. Against this supposed reality, cities are portrayed as beset by a kind of accumulation disorder and uncertainty that now haunts them. The city, as a place of becoming, is a threatened but historically resilient locale that once again must act ingenuously to survive. The offered signs of this ominous potentiality – municipal fiscal depletion, an aging physical infrastructure, the "reality" of decayed residential, commercial, and production spaces dotting the city – are deployed as disciplining signifiers of what the future can bring. Through this rhetoric, a proposed shock treatment of re-regulation and privatization is grounded and rationalized.

In a second part of the rhetoric, city survival supposedly depends upon following two imperatives: strengthening the city as a taut entrepreneurial space and meticulously containing black ghettos and their populations. In the first imperative, the assertion is forceful: Now cities must push to build attractive consumptive complexes, upper-income aesthetic residential spaces, efficient labor pools, and healthy business climates. This post-1990 rhetoric has been at the heart of what Kevin Cox (1993) earlier identified as the supplanting of a "politics of redistribution" by a "politics of resource attraction." Entertainment, culture, sports, and leisure now become civic business. To fail to commodify these, borrowing from Milwaukee Mayor J. Norquist (1998), is to miss the reality of the new stepped-up inter-city competition. An intensified fragmenting and balkanizing of city space by class and race is not merely normalized, it becomes celebrated as utilitarian and in the service of city survivability.

In the second imperative, the assertion is sometimes explicit but often implicit: that poor black neighborhoods and populations need to be systematically isolated and managed as tainted and civic-damaging outcasts. These are cast as not merely culturally problematic but things to be feared, reviled, and cordoned off. At work is William Wimsatt's (1998) notion of the mobilized fear economy, a general trepidation that now expands to more deeply include black ghettos. As Wimsatt notes, since 1980 we have increasingly had government by fear, foreign policy by fear, and landscapes of fear, all of which are expediently peddled by all scales of media. Now, we also have a heightened fear of the sinister black-ghetto in these cities that is manifested in a discursive fright about crime, black men, black youth, streets, and ghettos. A spiral of fear, peddled through rich images, now sells black bodies and spaces as potential violators of the collectivity's socio-moral and economic integrity. As is revealed in the analysis of contemporary ghetto changes (chapter 4), the unhidden hand of the global trope that sells this can be found in city policy, planning discourse, and normative politics.

The global trope is in this sense two-pronged. It offers the complementary "truths" of what circumstances these cities now face and also what they must do to survive. These two supportive formations seamlessly connect to form a coherent and resilient rhetoric. This whole, borrowing from Wendy Hollway (1984), offers purportedly progressive positions for subjects to adopt that legitimate potentially contentious actions (e.g. requiring poor people to work at sub-minimum wages, cutting food stamps to the needy, using public funds to subsidize gentrification). Yet use of such discourse by growth elites is anything but surprising. These formations, following Norman Fairclaugh (1992), are the modern alternative to flagrant violence and oppression. The now established rule in complex societies, to Fairclaugh, is to make and manage rather than to nakedly repress. To Fairclaugh, seizing and extending the terrain of logical and progressive through discourse, is potent politics.

The end result, I chronicle, has been the formation of a new kind of ghetto, what I term the "glocal black ghetto," which has become more impoverished and more impugned as the now crystallized zone of human discard in "the global era." These ghettos, simply put, have become one-dimensional apparatuses for the naked isolating and warehousing of those deemed cancerous to real-estate submarkets and downtown transformation. In the process, dominant changes in these ghettos (deepened deprivation, more health fatalities, new forms of stigma and marginalization) reflect this ghetto and inner city isolating imperative put into play. The facilitating rhetoric, the global trope, proves functional by communicating the need to re-entrepreneurialize city form and life and deepen ghetto isolation. Ultimately, it normalizes both an intensified splintering of city space and the sense of tainted and civic-damaging black outcast bodies that need assiduous regulating and management.

But use of this ghetto-devastating global trope in the third-wave is rooted in a deeper force that has so far been merely hinted at: the production of a strategic uneven development. This differentiation of city form has fluctuated over time in response to a central process: local and societal regimes of accumulation. This cultivating of uneven development, Neil Smith's (1984) lifeblood for making the city an instrument for accumulation, produces an economically-taut landscape that can efficiently service the interests of local growth machines and the broader society. Thus, during the golden age of the Fordist societal growth dynamic, rust belt cities like Chicago, Cleveland, St. Louis, and Detroit took on and progressively embellished their trademark feature: large factory districts dominating downtowns ringed by tiers of worker districts (Judd 1979; Teaford 1990). Black ghettos immediately arose to aid a small real-estate capital but most fundamentally to assist the Fordist industrial economy's need for cheap and plentiful low-wage workers.

But local and societal circumstances were changing in the 1970s with the collapse of Fordist economics and the Keynesian-welfarist complex. As flexible production systems, labor-market deregulation, and a retrenched welfare state became the societal adjustment, rust belt cities especially were battered. These cities, desperate to revitalize moribund economies, rallied around an "opportunity structure" provided by the structural economy, potentially lucrative real-estate (see Smith 2002), to drive the second-wave of black ghetto marginalization. Fluctuations in land and property value, as before, persisted, but cultivating an ascendant gentrification could generate substantial revenues for real-estate capital and local government (see Weber 2002; Smith 2002). In this context, the institutional stimulants to revalorize land in key districts – tourism, historic preservation, cultural upscaling – arose as city redevelopment mechanisms. Desires of growth machines to cultivate this new city-wide differentiation, steeped in isolating "contaminating" black bodies and building expansive (but fortressed) posh spaces, spurred the creation of the new glocal black ghetto.

AN UNEASY GLOBAL TROPE

Yet it is important to distinguish between the appearance and reality of these growth machines and their usage of the global trope. At a superficial level, they appear as blunt neoliberal operatives, flagrantly offering a kind of new shock treatment (e.g. necessity of concentrating public and private resources in select spaces, demanding the racial poor to be productive and civically contributory or pay the price). But things are more complex at a deeper level. These machines elaborately stage their power and acuity to appear as inevitable and irreversible forces (Pulido 2000). This "theater of self-aggrandizement" bolsters the machine's political standing and conceals the difficulties of its reality: it must continuously struggle to negotiate shifting political ground, engage new possibilities and constraints, and grapple with new forms of contestation (Ward 2000). If successful on these fronts, the myth of naturalness and inevitability is hardened and dogmatic and strident neoliberal rhetoric can proceed full force.

In this setting, the global trope is always multi-textured and elaborately staged to be effective and solvent. It "speaks" directly to specific issues (the reality of globalization and city need to appropriately respond) but fabricates elaborate worlds of people, places, and processes that foundationalize and organize these themes (see Wagner-Pacifici 1994; Castells 2004). This provision of "support worlds," a crucial analytic ingredient in the rhetoric, functions to stage these "themes of truth" as they connect to the lifeline of "truths" in other rhetorical formations. These support worlds, in other words, are necessary inclusions in the rhetoric that authenticate dominant, addressed issues. Mapping reality ultimately involves staging the mapping replete with providing a supportive cast of characters and processes. Thus, as we discover, the global trope's ability to persuade (i.e. create perceptions that make certain actions practical and others not) lies in a discursive framing of its dominant themes, which cultivates and manages the sense of one objective reality.

It follows that the global trope which drives this new uneven development is complex and tension-ridden. Contradictions and discontinuities characterize the formation – its themes, images, and general coherence – that need continuous management and refinement. This formation's complexity is tied to a straightforward reality: it is a strategy of power that is never complete or fully determinative. The global trope is thus always in a process of becoming, as something partial, contingent, and developing, to render it malleable, fluid, and hybridized. At the heart of this, the trope is always subjected to a "double-gaze," a two-sided observation and interpretation, which continuously opens it up for scrutiny and interrogation (see B. Wilson 2000). Young and old, the poor and non-poor, and everyone else take their turn at reading this formation. To dull or taint this gaze, the search for a consensus and the production of a democratic veneer is constant. Contestation and resistance, as we learn in chapters six and seven, is forever there or on the horizon, making the creation and reproduction of this global trope an ongoing human accomplishment.

What are the specifics of these difficulties? Most generally, a surprisingly elusive abstraction – new global times – is always being simplistically grounded and empiricized. The global trope is an elusive abstraction in a fundamental way. A sense of new global times is an absent reality, an empirical ambiguity (see Dear 2000; Cameron and Palen 2003). It is not visible to people in space, and is said to lie way beyond the domain of states and regions. It is also absent temporally, with globalization widely invoking the sense of an inexorable, futuristic unfolding as "the telos of capitalism." In this context, growth machines continuously toil to "proof" globalization as something observable, legible, and on the move. In this process, a sense of easy-to-understand local ills is widely served up as irrefutable evidence. Manifestations of globalization are projected to be all around the city: in people (e.g. the black poor), places (e.g. industrial districts), and processes (e.g. city crime, declining public revenues). The public is to see the city and quickly grasp this proof: the city is to be read in only one way.

The struggle is also to reinforce something else: the local state as leader of the new restructuring. To push this, growth machines extol the state's supposed reason for existence, to form and execute collective goals, even as prevailing neoliberal sensibilities also necessitate anti-statist rhetoric (see Ward 2000; Weber 2002). Direct pronouncements (government as facilitator of civic livability and civic progress) and subtle insinuation (government as preserver of status quo class and race relations) help these growth machines: they prop up this offering. In short, the push of a proactive government belies neoliberal orthodoxy. The drive to front a smart and adroit local state is a non-stop rhetorical project. Ultimately, these local states, in the growth realm, do not abandon (in action and discourse) sense of themselves as mechanical bearers of public desires that transform cities for public gain, even as they struggle with the new reality of having also to demonize themselves.

Moreover, these growth machines struggle with something else: they communicate the contradictory notions of democratic ideals and the need to isolate the black poor. While the principles of freedom and self-determination are extolled, policies blatantly isolate "a people." Rationalizing this confining, an ongoing project, involves a two-pronged process: bringing supportive, paralleling narratives into the global trope (e.g. the black crime question, the erosion of public schools issue) by referencing and illuminating; and allowing these narratives to function and influence on their own (see Pulido 2000). In theme, both offer a doctrine of liberty that is tied to a notion of deservedness to be measured by two supposed time-tested ideas: levels of civic conformity and civic contribution. In this context, poor African Americans are cast as a least deserving lot: they are widely demonized as threats to public safety, security, and civility (Hooks 1993; Collins 1996). Diverse discourses in the spheres of crime, public education, city growth, community development, and housing policy are critical. I discuss this more fully later.

At the same time, the agenda to isolate the black poor must be complete and total. This key part to creating the entrepreneurial-competitive city involves triple goals: the raw act of cordoning off "a people," rendering them accepting of this and non-incendiary; and removing totally their presence from the civic gaze onto privileged micro-spaces. Creating this new city becomes a delicate, ongoing human endeavor that involves deft discursive and material management. The final goal of these three (managing the civic gaze onto select micro-spaces) is perhaps most vexing; it necessitates a non-stop management of the black poor's activity spaces and routine paths. The growth machine's realization is stark: the images that these "cathedrals of consumption and production" emit need to be elaborately choreographed and controlled. It follows that such commodifying of space, goes hand-in-hand with a key maneuver, entrepreneurializing the visual and banishing "visual trash."

But who offers this global rhetoric in rust belt cities? The leaders are diverse "talking heads" within growth machines: planners, mayors, City Council people, newspaper writers, developers, Realtors, editorial pundits, and corporate CEOs. This is the nexus of enablers, funders, planners, writers, and direct builders of urban space who aspire to create a new, profit-propulsive capitalist city. They unify around a central goal: to produce maximum urban rent and to cash in on the produced revalorization of land. This means, of course, encouraging multiple city changes: attracting more business and industry; building more conspicuous consumption neighborhoods; crafting vibrant, lavish downtowns; re-entrepreneurializing local business climates, and isolating the racialized poor. These actors frequently differ in the desired timing and pattern of restructuring, but this fails to blunt the drive to restructure. All desire in general principle a coherent nexus of spaces that yields the prize: investment-attractive micro-terrains (e.g. gentrified neighborhoods, historic districts, high-tech production zones).

This combination speaks its truths through multiple sources: speeches, public oratory, newspaper editorials and stories, planning documents, and informal everyday conversations with colleagues and others. All help constitute a circuit of knowledge that permeates the urban everyday to populate the local with anointed facts and realisms designated as irrefutable (hence this book's empirical focus on all of these sources). One key point is the "regime of truth", which is dependent upon a crucial but often overlooked source – the mundane everyday conversations of growth machine actors, as bold declarative statements in public forums. It continuously replenishes as a foundational source of global-speak, the content and legitimacy of the discursive formation as neoliberal infused ideas seamlessly pass from one actor (growth machine voices) to others (both growth machine actors and others), albeit in informal settings. Such everyday conversations, ultimately key builders of truth for growth machine members and residents alike, are active at every moment in the circuit's life.

In this context, these renditions, as meticulously set-up and ensnaring worlds, feature seductive, prominently haunting images, which draw people into their stocks of truths. One prime image, for example, conveys an entrepreneurially robust, aesthetically and culturally dreamy city easily made with strong public support. People, through this, are taken along imagined paths resonating with adroit symbols and indicators of civic prospects and potentialities. Another common image, its relational other, presents something very different, a currently threatened, de-stabilized city in new global times. The proof is stamped into the entirety of the image in the form of boarded-up storefronts, sinking shops and retail zones, crumbling neighborhoods, malaised downtowns, failing schools and rising crime. Wherever one looks in these staged images, self-fulfilling signs of an uncompromising and harsh global economy lurk. Not surprisingly, any potential evidence that contradicts the offered "reality" is purged from the images.

A key implication arises from recognizing the reality of this deployed global trope: it refutes the near mantra-like belief that city growth machines and "globalization" are inherently oppositional forces. This recognition thus suggests that it is false to automatically counterpose city growth machines and "globalization" as antithetical. Globalization, many analysts declare, is a mobility-enabling force for capital that necessarily runs counter to city health and growth machine designs. A process termed globalization, at every second, is seen to assault the desperate growth machine imperative to keep cities robust and vibrant. For example, Holston and Apparadurai (2003) capture this thematic, noting that "recent developments in the globalization of capital . . . drive a deeper wedge between national space and its urban centers."

This study suggests something different: that these machines, as centers of rhetorical production and power, can seize the day's concerns and constitute and reconstitute a sense of powerful globalization, which helps their restructuring ambitions. Globalization, as a served-up construction, bolsters a fervent desire of growth machines: to ensnare "trophy investments" and restructure cities to their specifications (Zukin 1995). An invoked reality of globalization, in this sense, helps foundationalize and expand a neoliberal social and physical restructuring that is at the core of current growth machine aspirations. To be sure, these coalitions have not invented this global concept, and it does exist in reality as an elusive, highly uneven process, but they continue to draw on it, magnify it, and caricature it as they take advantage of this ambiguous notion in common thought.

In this context, diverse cities in America's rust belt are examined: Chicago, St. Louis, Indianapolis, Cleveland, Detroit, Pittsburgh, New York, Philadelphia, and others. These rust belt cities cover a vast stretch of terrain that runs from Minnesota on the north, Missouri on the south, the northeastern seaboard on the east, and the Mississippi River on the west. These cities and their ghettos share the legacy of having economic and political bases rooted in smokestack manufacturing that dominated America's nineteenth and twentieth century industrial might. Chicago's rootedness in meatpacking and steel production, Detroit in automotive manufacturing, Pittsburgh in steel production, and Philadelphia in metals were but the leading edge of a once massive complex of heavy-duty production that structured the organization of neighborhoods, industrial districts, social spaces, and social relations.

Today, amid all the tumultuousness and change, these cities, following Amin (2002), are still not places that are "nested in simple territorial or geometric space." They are, rather, "nodes in relational settings . . . locations of situated practice[s] . . . a place of engagement" where history, power, and practices collide to forge distinctive arenas for human action. These black ghettos, like their encompassing cities, are thus different: Cleveland's Hough is not Philadelphia's Fairhill, Chicago's Wentworth differs from Indianapolis's Eastside. Forces that affect them are rooted in place-specific cultural histories, political cultures, and complex institutional climates. In this sense, these ghettos are constituted through different place-based institutions, social fabrics, and political-regulative formations. But these ghettos, I argue, share the key commonality of being historic storehouses for the poorest African Americans. Currently, they are all profoundly affected by the latest broad-sweeping assertions of the new global era in deepened conservative times.

PERSPECTIVE AND DEFINITIONS

This work uses what I call a racial economy perspective to deepen understanding of these ghettos. It draws on the now well-known schools of urban political economy and racial-cultural studies to understand the evolution of these spaces and their current dimensions (see Lott 1999; Pulido 2000). The goal, as in so many post-structuralist studies, is to recognize the importance of three analytic spheres as they condition life and are lived through – race, economy, and culture. Yet these "spheres" are seen to have anything but clear and easily delineated boundaries. I thus suggest that "race," "culture," and "economy" exist in rust belt cities not as empirically separate things, but as inseparable, nested elements in power-laden social formations. These nesting constructions, as lived arenas for people, are meaningful to growth machines: they can be constructed and drawn on to wield power and influence as inputs into regulatory formations.

The core of the racial economy perspective is a belief that a humanly produced element, race, has intimate ties to politically-infused economies in places. Producing and working through race – "racializing" the everyday – is a practical and technical accomplishment that helps fix and maintain social relations to the material and symbolic benefit of some. Production of race is ultimately compelled more by real interests and discursive strategies than by attempts at factual, real-world reportage. Race, in this sense, is not only a social construction, but also a key cog in an elaborate circuitry of power. Its construction, seizure, and usage lubricates the economic machinery of daily life. Yet race is more than simple ascription: it is a constitution of regimes of images and relations of meaning that help colonize the common vision about places, people, and processes. Through producing race, then, power is provided a "realness" and legitimacy that links racialization to the nub of the everyday's ritualized thought, social practice, and common conduct.

Space is also important in this racial economy perspective. Space, borrowing from Brenner and Theodore (2002), is now one privileged instrument through which racial economy operates. Most immediately, processes framed at meshing spatial scales – the local, the regional, and the national – enable racial economies to forge the likes of this study's central analytic object: black ghettos. Scale, here, discursively stages the world that, in offering one expanse of reality, imbues presentations of forces and processes with credibility. One discrete "visual" of the world is set out and sedimented to privilege the existence of certain forces and processes. Banished to oblivion, in the process, are the referent of other scales and their power to lend credence to the realness of other processes. To dictate scale, following Livingstone (1992), is to wield cultural power. Scale, then, can fruitfully be seen as a kind of resource, albeit a human made one, whose strategic usage propels racial economies forward.

But what is meant by black ghetto in this study? I mean to identify a socially isolated, segregated class-race space that today more staunchly isolates poor

African Americans as growth machines struggle to make a differentiated city space. I thus refer to a terrain of neglect and ethno-confinement that is put in the service of a dominant status group. Extending the view of Wacquant (2002), this ghetto is a socio-spatial device that enables a dominant social group in cities to ostracize and exploit a subordinate group endowed with negative symbolic capital. This relation of ethno-racial management and closure involves four aspects: stigma, constraint, territorial confinement, and institutional encasement. This ghetto's daily rhythms, as a distinctive entity, follow from the four African-American ghetto stages in America (slavery, Jim Crow, the incipient ghetto, the hyperghetto) identified by Wacquant (2002) (Table 1). More than before, in this black ghetto, "a people" are socially and spatially cordoned off from the mainstream as supposed contaminants to the public good.

These ghettos are defined, for operational purposes, as spaces with more than 95 percent of residents African American and with 35 percent or more of households living below the poverty level. These communities also had to be within the city's political boundaries and not be a separate, incorporated municipality. These numbers, adopted to capture critical assemblages of this local racializing and impoverishment, allow us to include many of the classic black ghettos studied by others (e.g. Hough in Cleveland, Bedford-Stuyvesant and the South Bronx in New York, Wentworth and Woodlawn in Chicago, and Allegheny West and Hartranft in Philadelphia). Recent definitions by Jargowsky (1997, 2002) and Petit and Kingsley (2003) are similar (in studies of "extreme-poverty neighborhoods"), but set this poverty figure at the slightly higher rate of 40 percent.

What proof is there that these black ghettos have recently worsened and been functionally re-cast since 1990? Most immediately, the data that is difficult

Table 1 The nature of the now five black poor "peculiar institutions"

Institution	*location*	*form of labor*	*core of economy*
Slavery 1619–1865	regional south	unfree fixed labor	plantation
Jim Crow (1865–1965)	regional south	free fixed labor	agrarian and extractive
ghetto (1880–1968)	U.S. cities	low-wage industrial	menial worker
hyperghetto and prison (1968–1990)	U.S. inner cities	residual postindustrial services	marginal, service-oriented
glocal ghetto (1990–)	U.S. inner cities	underground fixed, forced statist	underground, shadow and marginal, service-oriented

Source: derived partially from Wacquant (2002)

Table 2 Changes in black ghetto neighborhoods[1]

	% Population Change, 1990–2000	*% Below Poverty Level, 2000*	*% Below Poverty Level Change 1990–*	*% of Housing Units Vacant, 2000*	*% Change In Housing Units Vacant, 1990–2000*	*% Change Ratio of High to Low Poverty 1990–2000*[2]
Cleveland						
Fairfax	−13.1	35.7	−0.9	21.0	+3.6	+24.2
Hough	−19.2	41.3	−2.3	20.9	+0.6	+41.3
Philadelphia						
Fairhill	−22.8	57.1	+2.0	22.0	+5.8	+31.3
Hartranft	−7.0	33.9	+1.4	21.8	+7.8	+18.7
Chicago						
Englewood Woodlawn	−16.8	43.8	+2.9	23.7	+8.7	+22.9
Baltimore						
Boyd Booth	−7.6	38.3	+1.2	26.7	+9.0	+21.2
Broadway East	−11.2	39.0	+2.4	17.4	+11.7	+12.6
Detroit						
Planning Cluster I	−30.3	38.0	+0.5	13.1	+2.8	+27.5
Planning Cluster II	−20.0	36.3	−1.1	16.7	+5.1	+31.3
Washington						
Trinidad	−7.1	41.3	+1.8	18.8	+11.1	+19.7
Bellevue	−6.8	40.9	+0.3	17.6	+14.7	+23.9

[1] Unit of analysis for computation: census tract
[2] high poverty measured by people with incomes two times or greater below the poverty level. Low poverty measured by people with incomes .50 or less below the poverty level.

to refute. First, material worsening is shown by data from a sample of these ghettos across the rust belt (Table 2). Eight of the eleven neighborhoods explored in the six cities experienced increases in families living below the poverty level (an average increase of 0.8 percent), the percentage of housing units vacant (an average increase of 5.5 percent), and poverty populations that were "high poverty" between 1990 and 2000 (an average increase of 24.9 percent). All eleven neighborhoods also experienced substantial losses in population, with two areas, Planning Cluster I and Fairhill in Detroit and Cleveland, losing more than 20 percent of their populations. In sum, all eleven randomly selected poverty neighborhoods fared worse in 2000 than in 1990 on all four variables.

With further review, some of the numbers are frightful. Cleveland's Hough and Fairfax experienced increases of 41.3 percent and 24.2 percent in their ratios of high poverty to low poverty residents. This index is especially revealing, measuring change in the intensity of deprivation within poverty populations over time. Philadelphia's Fairhill and Hartranft had, respectively, 57.1 percent and 33.9 percent of their populations officially living below the poverty level, a 2.0 percent and 1.4 percent increase, respectively, from ten years earlier. Chicago's Englewood similarly had an official poverty rate above 43 percent. Planning Clusters I and II in Detroit suffered equally experiencing growth in high poverty residents as a ratio of low poverty residents of 27.5 percent and 31.3 percent, respectively. The data is unequivocal: in poverty-afflicted and dilapidated inner cities that were battered in the 1990s, these areas suffered the worst.

To be sure, some of the forces that assault these ghettos also afflict other working-class populations in the rust belt and beyond. Thus, this ghetto population anchors the newest grim statistics about growing despair and poverty in America. For example, the number of Americans living below the poverty line increased by more than 3.5 million from 2000 to 2002 (to 34.6 million) (*Chicago Tribune* 2004). In a similar statistic, those who are unclear where their next meal will come from, termed "food insecure" by the U.S. Department of Agriculture, grew from 31 million to 35 million between 1999 and 2002 (*Chicago Tribune* 2004). In 1970, 4 million people sought food assistance through food stamps; in 2003, the figure was 23.5 million people. But this poverty has been concentrated in an anything-but-surprising place – these ghettos, where residents are largely low-income, struggle to negotiate the new urban service economy, and are powerfully stigmatized. Most vulnerable economically, a tripartite of race, class, and stigmatized setting entraps and punishes a population.

Descriptive accounts of living conditions in these rust belt ghettos from writers across the political spectrum bolster this notion of deepened deprivation. To *Detroit Free Press* columnist Fred Payne (2002), Detroit's poorest black neighborhoods seem more ravaged and neglected than ever. To Payne, this "zone" now has but one movie theater and a few retail stores. To find a Sears or a Marshall's, these residents have to travel to neighboring Dearborn. The nearest fast food places, Popeye's and McDonalds on the main drag Woodward Avenue, serve food from behind bulletproof glass. The city's unbroken rows of abandoned buildings, an estimated 10,700, cluster in these ghettos. Nearly 1,200 of them are found within one block of inner city public schools (*Detroit Evening News* 2001). The Riverside neighborhood, one of the city's most impoverished, had one-fourth of its housing stock (222 buildings) ravaged by abandonment in 2001 (*Detroit Evening News* 2001).

Chicago's black ghettos are similarly described. Urban League writer Paul Street (2003) finds despair in six Chicago neighborhoods where more than 40 percent of kids are "deeply poor" – Oakland, North Lawndale, Washington Park, Grand Boulevard, Douglas, and Riverdale. Unrelenting hunger, homelessness, and drug abuse, Street reports, punctuate the streets and parks of these communities. As noted by W. J. Wilson (1996), vacant land and abandoned buildings from general institutional withdrawal punctuate this physical fabric. To exacerbate the community's stigma, roughly 60 of the city's

80 recently installed surveillance cameras now dot community "hot spots" (Clarke 2004). Operation Disruption Surveillance, initiated in 2003, has spent $3.5 million to detect street crime and monitor the activities of Chicago residents (see *Chicago Tribune* 2004). Now, these kids and adults are constantly watched in the city's proclaimed "blue-light districts" (Chicago Planner D. Roe 2004). This "soft" use of electronic surveillance, imperceptible and harmless to outsiders, reinforces the criminalizing of a population.

Evidence also suggests that these black neighborhoods have been representationally re-cast as more culturally and civically problematic spaces since 1990. First, proof comes from the media with the range of its reportage-types about ghettos considered (i.e. editorials, community exposés, crime reporting). Data from a sample of four daily newspapers in rust belt cities, the *Cleveland Plain Dealer*, *St. Louis Post Dispatch*, *Indianapolis Star*, and *Detroit Free Press*, shows a more frequent reporting of black ghettos via use of a negative metaphoricalizing, as pathologically consumptive, after 1990 compared to the mid 1980s (Table 3). This differed from the common media and city rhetoric in the previous period, 1980s Reagan era, that emphasized something equally inciting but less "complete:" a dramatically falling-into-pathology population in these spaces (Wilson 2005). Whereas articles in the 1980s widely reported an incendiary process, reports in the 1990s often chronicled the reality of a complete downward spiral. The most flagrant example of the latter, from the *Indianapolis Star*, had a reportage increase from 6 percent in 1985–91 to 14 and 12 percent in 1992–97 and 1998–2003, respectively.

Second, discussions with local planners and politicians in the cities underscored this representational re-casting of black ghettos. These people, also, frequently referenced or discussed their city from the position of these residents and spaces as civically non-contributory and unproductive. The dynamic at work was a kind of "deeper slide into normalcy" (in planner and politician common thought and practice) of warehousing poor black families, an okaying and sanctioning of segregation. Two kinds of response reflected this. First, discussions of ideal residential structure across these cities produced little commentary on the reality and ills of segregating the racialized poor. This was all-but-off the planning agenda, supplanted by such concerns as "Smart Growth" and "the New Urbanism." Second, those that discussed poor black neighborhoods often centered their function within the notion

Table 3 Percent of stories presenting black ghettos as pathologically consumptive and obstacles to city growth

	1985–1991	*1992–1997*	*1998–2003*
Cleveland Plain Dealer	1 (2%)	5 (10%)	6 (12%)
St. Louis Post Dispatch	1 (2%)	6 (12%)	7 (14%)
Indianapolis Star	3 (6%)	7 (14%)	6 (12%)
Detroit Free Press	2 (4%)	5 (10%)	6 (12%)

of needing to cultivate the broader city (as a supposed fragile economic and social landscape in a new global era). To many of these planners and politicians, who spoke of themselves as civic servants, this was the public's purported central concern less so poverty, deprivation, or anything else. Two quotes capture the essence of these two responses:

> What the public wants in Chicago is livable, usable spaces. That is why the new urbanism has a large following here. The unit of importance is the neighborhood, and Chicago is a city of neighborhoods. Our planning goal . . . is to make this a reality, build a city that the people want and can thrive in.
>
> (Chicago Planner B. Walters 2004)

> Black poverty still plagues the city, it's found too frequently . . . It's admittedly a tough situation, welfare doesn't meet their needs and desires . . . the workfare experiment seems to be working . . . St. Louis is becoming a national symbol of urban recovery and progress, these neighborhoods at best don't help the process . . . at the worst, they hinder it . . . They need to play a more productive role in the St. Louis economy.
>
> (St. Louis Planner M. Wilks 2005)

A note on the methods used in this study. Textual analysis, open-ended discussions, and content analysis of a radio talk show were the data sources. Textual analysis deconstructed stories about city growth and redevelopment in seven local dailies (e.g. *Chicago Tribune, Chicago Sun-Times, Cleveland Plain-Dealer, Detroit News, Indianapolis Star, St. Louis Post-Dispatch, New York Times*) and on the web. Stories and articles using the terms growth, redevelopment, globalization, or ghetto were identified for review. Open-ended discussions were also conducted in six rust belt cities in 2004 and 2005 – Chicago, Indianapolis, Cleveland, Philadelphia, St. Louis, and New York. I conversed with local planners, city officials, city program heads and representatives, community activists, residents, and youth in person or by telephone. To obtain credible responses, all interviewees were initially asked if they preferred to have their names withheld from future write-ups of the data. Nearly 90 percent of the 130 interviewees opted for this. For this reason, comments by discussants in the book frequently fail to carry a name or simply provide a pseudonym.

A final source of data was a content analysis of the nationally syndicated Mancow Muller radio talk show. This text was ideal for capturing the pulse of current political thought in the neoliberal-infused rust belt. His frequent diatribes about the black poor, black ghettos, the welfare state, new global times, and the politics of racism were resonant and revealing, reflecting the ascendancy of the deepened post-1980 conservatism. His oratory, often deliberately invoking incendiary images, nevertheless spoke about deeply felt beliefs. It is no accident that Muller is now carried on over 25 radio stations across America and his ratings are booming (he was *Billboard Magazine*'s Radio Personality of the Year in 1995, 1996, and 1997). Muller, now established at the center of America's growing list of neo-conservative talking heads on T.V. and radio, reflects and fashions mainstream political beliefs.

A final brief comment of self-reflection. Is this book an unequivocal presentation of truth? I believe yes and no. I borrow Michael Keith's (1993) pronouncement that any academic work is unavoidably a relativist human-made product, a kind of situated, cerebral output that we affix as a thing called knowledge. This product is socially constructed through discursive formations that arrive at truths through the unavoidable use of language, political perspective, and cultural meanings. Keith terms this producing of knowledge "hard labour at the coalface." This book thus speaks its truths through this degree of relativism, but I believe these to be ultimately valuable. Thus, this work is seen to open up a kind of aperture to see rust belt cities and their black ghettos in a distinctive way: through a racial economy perspective. In this work's inevitable imposition of cultural meanings and obliterations, ontological presences and absences, and linguistic tropes, one important reality is promulgated for others to see and reflect on.

2 Rise of glocal ghetto

THE BEGINNING

While people live and breathe in these rust belt cities and ghettos, the brute functionality imposed on them in capitalist America is difficult to refute. A continuous tension between age-old urban adversaries, those who drive to repetitiously functionalize the city's physical and social order and those who strive for livable, affordable communities, has been at the heart of defining urban form and social space (Ferman 1996; Jonas and Wilson 1999). This chapter works from this conceptual position to understand the evolution of these cities and black ghettos over time. As is documented, the newest, post-1990 form of this, Gordon MacLeod's (2002) splintered post-industrial city, frequently involves producing and deploying black ghettos as iron-fisted store-houses to help establish the latest "post-rust" inclusions: high-tech zones, expanding gentrified landscapes, high-culture public spaces, and conspicuous consumption retail corridors. A steering and isolating of "contaminating" black bodies to their own dead-end universes, in this dynamic, has been un-relenting and further institutionalized.

In this context, modern black ghettos in rustbelt cities began in the 1920s and 1930s with the Great Black Migration and the drive of urban elites to amass and control cheap labor. Booming industrialization in the rust belt and displacement of labor from southern farms compelled rural blacks to enter the northern Fordist industrial economy. In 1910, 92 percent of blacks in America lived in the south. In 1950, 65 percent of this population lived in the north and east (Forman 1971). Strongest initial magnets for these migrants were Detroit, Cleveland, New York, Baltimore, and Chicago. For example, Detroit's and Chicago's black population went from 6,000 and 44,000, respectively, in 1910 to 120,000 and 235,000, respectively, in 1940 (Johnson 2003). The creation of old or first-generation ghettos in these cities (Rose 1971), defined as poor black enclaves with 25,000 or more people before 1920, followed that built on pre-1910 migration streams of southern blacks.

When blacks moved into these ghettos they were often surprised to find intense racism and discrimination. Southern white values about race, they discovered, were also prevalent in the north. For example, many restaurants and stores in Indianapolis, Chicago, and St. Louis refused to serve blacks (see Osofsky 1967). At the same time, parks, cemeteries, beaches, and hospitals were typically divided into "white" and "colored" sections, and unhealthy living conditions were the norm (see Speare 1969). In Detroit, between 1920 and 1930, arrest rates for blacks were four times that for whites (Judd 1979). Blacks, only 9.3 percent of the nation's total population in 1930, constituted 31.3 percent of the prison population. In New York's Harlem, between 1923 and 1927, the then youthful population had a death rate that was 42 percent higher than the city. This area's infant mortality rate, moreover, was 111 per 1,000 births compared to the city's 64 per 1,000 (Judd 1979).

By 1940, with accelerating industrialization and in-migration, black ghettos had grown in these cities. Philadelphia's North Side (175,000), Chicago's Black Belt (190,000 people), and New York's Harlem (200,000 people) grew by more than 300 percent in twenty years. But second-generation ghettos also appeared at this time in Harold Rose's (1971) second-tier industrial cities: Cincinnati, Boston, Indianapolis, Cleveland, Newark, and St. Louis (see also Still 1974; Teaford 1990). Ghettos in these cities, exceeding 25,000 between 1920 and 1945, included most notoriously Hough and Glenville (Cleveland), Russell Woods (Detroit), North Ward (Newark), and Inner City (St. Louis). Like their first-generation counterparts, industrialization dramatically swept over these cities that attracted new immigrants from the rural south. Frequently unable to find housing outside established ghetto enclaves, these second-generation ghettos grew quickly (see Teaford 1990).

First- and second-generation ghettos became more isolated and stigmatized by the late 1940s (c.f. Drake and Cayton 1945; Osofsky 1967). At the core was the rise of a dominant, low-wage laborer class and segregated institutional networks in these areas (black churches, black press, black stores, black fraternal lodges). A fully emergent "ghetto-within-a-city" evolved that was a startlingly successful capitalist creation: cities could efficiently extract black labor while "the civic contaminating influence" of black bodies could be isolated. Black bodies, tactically distributed in space and assiduously controlled, were now more fully objects of municipal surveillance, control, and spatial management. Residency in these spaces meant the simultaneity of being marked as low-skilled and undesirable, living in substandard housing, going to separate and unequal schools, and working in factories at miniscule wages to build wealth for others (see Wacquant 2002a). The rust belt black ghetto's modern form had emerged, which would persist to the present.

But these 1940s ghettos were also constituted and re-constituted by key institutional practices. Two processes were crucial – zoning and realtor steering – that were to haunt these cities thereafter. These classic institutional instruments of the Fordist age came to the fore after 1930. Each helped to re-make and piece together a new city, a robust industrial center, which applied state and capital's power through embedded routines and a sense of efficient protocol. Its fulcrum was the use of technical experts (planners, builders, Realtors) and the de facto promise of civic-assisting acts. However, always securing the legitimacy to continue these practices was unpredictable. As Habermas (1973) notes, extending the state to support a biased market instantaneously exposes the political and legal grounding of capital's legitimacy, which can then be openly interrogated and resisted (see also Weber 2002). Thus, as zoning and steering were widely used across rust belt cities, their political base became widely known and periodically contested via protests and lawsuits (see Teaford 1990). Promising a mastery over the irrational and ambiguous, they sometimes became unexpectedly unstable and vulnerable.

Zoning, initiated in New York City in 1916, spread rapidly with a controversial constitutional confirmation in 1926. The U.S. Supreme Court ruled for the right of cities to subsume the unfettered individual's ability to determine use of land in favor of constricting this to plan for the general public interest. By 1930, 768 municipalities, with 60 percent of the nation's urban population, had zoning ordinances (Judd 1979). Realtor steering, too, marked the majority of rust belt cities by 1940 (Sugrue 1993). Steering by Realtors and confining poor blacks to ghettos became ritualized practices in Milwaukee, Chicago, Detroit, Cleveland, Cincinnati, New York, Pittsburgh, and Philadelphia (see for example, Kusmer 1976; Sugrue 1993). Land and housing markets in these cities during this period, experienced stepped-up control and regulation that would supposedly prevent a traumatic development: the unfolding of a chaotic urban form and a problematic infrastructure for city growth (Boyer 1983).

Zoning was initiated in a dramatically industrialized and crowded New York City. Here and elsewhere, it had at its heart the making of privileged city sections (the downtown, massive industrial districts) and turning them into actual and symbolic stages for economic clout (Boyer 1983). Its logic in New York was thus steeped in two drives: to feed this emergent industrial giant and protect real-estate values. By 1940, New York had more than 10,000 industries, more than any U.S. city and roughly 11 percent of the U.S. total (Page 2001). As support, zoning staked out a logical order of industrial districts, working-class neighborhoods, and warehouse zones. New York business and planning elites at this time imagined and pursued a vision of a preeminent economic giant (Caro 1974). At the same time, real-estate interests sought to protect Manhattan's elite neighborhoods, particularly its posh Upper West Side. This district, real-estate interests realized, had been displaced after a series of "invasions" of lower value land, which could happen again.

Zoning ordinances spread across rust belt cities in the 1930s and 1940s. Zones with specified densities and particular kinds of development chopped up urban form into sets of specialized and functionally serving fragments to render an optimal whole. This tool created more spatially disciplined cities to drive (as in New York) the two key wealth-creating apparatuses at that time: industrial and real-estate capital accumulation. Industries needed coherent groupings of neighborhoods and districts to make production efficient; real-estate interests desired balkanized "islands of neighborhoods" to cultivate healthy property submarkets. By 1945, with zoning, Milwaukee, Chicago,

Cleveland, St. Louis, Philadelphia, and Detroit had forged gigantic downtown industrial districts ringed by "feeder" labor neighborhoods (see Miller 1973; Teaford 1990). At the same time, elite residential areas were being buttressed by supportive large-lot zoning that minimized possibilities for "invasion" by low value land (Teaford 1990).

These zoning ordinances, not surprisingly, mapped out black ghettos for regulation. The usual recipe was to allow all kinds of land-use, small lot sizes for new development, and use of any building materials. A scheme evolved for managing these zones: housing for the racialized poor would be provided, poor blacks would live in their own worlds and not contaminate other housing submarkets, and their labor would feed the burgeoning industry of the cities. Now blacks disappeared behind walls fencing "a people" off in everyday city life. City zoning, as a whole, confined activity spaces and determined physical forms in these neighborhoods that helped engineer a marginalized population. Segregation and marginalization of the black poor, in a strategic stroke, emerged from the direct application of government resources and power. The interconnection of key things, space-making, social engineering, and government largesse, ultimately propelled this isolationist project forward throughout these years.

Atlanta's zoning ordinance most flagrantly embodied segregationist principles in this era (see Boyer 1983). Residential areas were divided into three racial districts: all white, all black, and undetermined. It was officially illegal for blacks and whites to live in each other's districts. This intervention into land market dynamics received mixed reviews, even from the growth and economic elite. Some opposed this, mainly because of the unprecedented involvement of the public sector. Others in this sphere supported it, seeing a powerful apparatus to help mold a sense of orderly housing submarkets. At issue was the conflicting desire to control the local state's police power but to use government to produce and insulate housing submarkets. This tension was ultimately resolved in an anything-but surprising way: zoning was accepted as a planning tool but amid stepped-up oratory about the need to be watchful of and control government influence (Boyer 1983). As in other U.S. cities, zoning quickly captured and assumed the status of political normalcy. Spurred by this regulation, by 1950, Atlanta's most pernicious black ghetto, Techwood–North Avenue, sprawled across its inner city.

Indianapolis's East Side was zoned as the ultimate undesirable area. It was configured to warehouse not only poor blacks but also noxious industries, flophouses, refuse facilities, and marginal retailing (Wilson 1993). Local planners and politicians in this era informally called the area "the dump", which persists as a moniker today (Howard 1991). At the same time, Detroit's double-barreled instruments of zoning and racial covenants proved extraordinarily successful in consolidating their black ghettos. Use of low-lot, "anything acceptable" zoning in its inner city, with the proliferation of racial covenants in deeds of homes outside this area, helped create a classic dual city. Creation of separate worlds for shopping, recreation, and living was inevitable and, in Detroit, powerful (see Herron 1993). To Herron, separate and unequal worlds in Detroit after 1945 could be (and continue to be) lived for days, months, and years with shockingly minimal contact across lines of fracture.

Through the 1930s and 1940s, Realtor steering was equally responsible for consolidating these ghettos. Its influence in these cities was sporadic prior to 1920: black neighborhoods were small and stigma informally segregated new black households (Osofsky 1967). But with the rapid influx of blacks into these cities, Realtor steering increased. Institutionalizing this, the National Association of Real Estate Boards in 1924 offered the stunningly explicit article 34 in their national code of ethics: "a Realtor should never be instrumental in introducing into a neighborhood . . . members for any race or nationality . . . whose presence will clearly be detrimental to property values in that neighborhood." A powerful institution, one that exerted profound control on the creation of the urban residential mosaic, in effect advocated the establishing of "poor black zones."

The drive for profit motivated Realtors to steer: it created a segregated residential mosaic that enhanced profits from sales commissions. But an elaborate ideology in this era gave the practice a normalcy and sanitization. Many Realtors contended that blacks lowered moral standing and property values in all areas (see Helper 1969). These Realtors perpetuated a common belief in this era, still prevalent today, that race, culture, and morals are intimately bound (see Balibar and Wallerstein 1994). For example, Helper's (1969) interviews with Realtors in this era discovered this ideology built around cultural differences and religious beliefs. Agents in Chicago, the study area, often invoked God, the U.S. Constitution, or irreversible cultural predilections of races as the source for their restrictive practices. Many Realtors articulated one of three themes: God did not intend the races to mingle and thus made them distinctive, the Constitution allowed people to segregate if they wanted, or different values of blacks and whites dictated a "logical" separating.

Not surprisingly, then, Realtor steering was both subtle and blatant. On the blatant side, for example, Milwaukee's powerful Real Estate Board declared to the public that . . . "the Negro population of the city is growing rapidly [and] something will have to be done" (in Forman 1971). The need, the Board concluded to the public, was to foster "a Black Belt." On the subtle side for example. Realtors in Buffalo pervasively discouraged out-migration from the city's ascendant North Side black ghetto, thereby reproducing its compostion. Black families outside this zone were frequently told that properties were rented or large utility bills were the reality not mentioned in ads. These families, as an alternative, were often referred to low-income housing projects. In Cleveland, the emerging Hough ghetto was reinforced by similar misrepresentations (see Hirsch 1976). Thus, vacant houses were often said to be sold or they were purportedly transferred to other real-estate companies, preventing purchase. At the same time, house prices were frequently marked up when black families made inquiries about the possibility of purchase.

The notorious National Association of Real Estate Boards (NAREB) spearheaded these actions. This Realtor boldly presented itself as "sensibly conservative" and "anti-welfarist." Its then long-term vice-president, Herbert U. Nelson, proclaimed himself a defender of free enterprise and a critic of anti-capitalist practices (Judd 1979). Under him, NAREB described itself as "a trade and professional association, [working] to improve the real estate business, to exchange information and, incidentally, to seek to protect the commodity in which we deal, which is real property, and to make home ownership and the ownership of property both desirable and secure" (in Judd 1979). NAREB, in press releases, commonly railed against repetitiously identified demons, segregation-busting social policy in cities, public housing, and anti-zoning legal actions, which were said to be destructive to the civic good (Gelfand 1965). What they advocated, the systematic cultivation of balkanized cities by race and class, was supposedly a recipe for urban order and social coherence.

The Federal Influence: 1940–1965[3]

Two federal government programs after 1940, public housing and urban renewal, also emerged to solidify the character of these black ghettos. Both can fruitfully be seen as post-war institutional fixes onto already blighted and deteriorating industrial cities (Boyer 1983). At the same time, black in-migration to these cities from the rural south was accelerating, which dramatically expanded this population and the size of their neighborhoods. In the 1950s alone, black populations increased 13, 17, and 23 percent in Chicago, St. Louis, and Detroit, respectively (U.S. Census Bureau 1950 and 1960). Growth machines, envisioning once finely articulated spatial divisions of labor and corresponding activity spaces, not surprisingly reacted (see Beauregard 1993). An institutional response to "reclaim" the early Fordist city structure was called for, the dilemma that Mayor Hubert Humphrey of Minneapolis (1948) called "the ulcer [that] may develop into the cancer . . . that [can] eat up our revenues and destroy our strength . . ."

Public housing and urban renewal gained legitimacy in rust belt cities as articulated policy thrusts embedded within a luminous icon – the comprehensive plan (see Swyngedouw, Moulaert, and Rodriguez 2002). Policy through the plan acquired acceptance and became the central mechanism for planning and restructuring through grand strategies in the early twentieth century (e.g. the city beautiful movement, the livable cities drive of the 1940s) (see Haworth 1966; Melosi 1981). This plan approach, intermixing notions of planning technocratic expertise, a unified wholism, attention to detail and morphology, and flexible re-working of space, was extended throughout the 1960s through public housing and urban renewal. Both initiatives, in this sense, clothed themselves in rich symbolism to convey notions of innovation, creativity, and spontaneity. The classic policy thrust of the Fordist age – the comprehensive plan – continued to prove resilient in rust belt cities.

Public housing had been initiated on a small scale in 1934; less than 20 U.S. cities contained this in 1940 (Wilson 1966). However, after 1955, the program was given a new urgency with the growing sense among city growth machines that cities were spatially hemorrhaging and "minority neighborhoods" needed to be contained. Between 1949 and 1967, more than 600 public housing projects were launched in some 700 cities (Jakle and Wilson 1992). Over half were to house more than 500 families. In planning schemes, central portions of downtowns were targeted for destruction and public housing construction. Many projects were designed to capture displacees from nearby urban renewal projects, others were envisioned as new neighborhoods for the poor (Wilson 1966). By 1970, over 450 public housing projects in U.S. cities had been built. Typically, projects were monstrous in scale (usually over 10 acres in size and housing thousands of people) and with a horrifyingly high concentration of people.

The effects on black ghettos were devastating. These projects flagrantly isolated and stigmatized black residents to a degree that embarrassed even some conservative politicians. Chicago's Robert Taylor Homes and Stateway Gardens packed 26,000 and 6,900 people in 28 buildings and 8 buildings, respectively. Located in Chicago's sprawling South Side Black Belt, this area became Devira Beverly's (1991) "world unto itself." The Robert Taylor Homes, 4,400 units in 28 identical 16-story buildings, resembled a prison set off from the normal world. Cages of meshed wire, encircling the buildings, guarded a seemingly incarcerated people. Indianapolis's Hawthorne Place Apartments and Concord Village placed 2,000 and 1,500 people on seven acres of land. Placed in Indy's "forgotten East Side," ghetto conditions deteriorated. Detroit's Charles Terrace Complex and Jeffries Homes collectively stashed nearly 5,000 people on nine acres of enclosed land; a population density of more than 500 people per square mile was closer to a third world city than the Detroit average (see Herron 1993).

But this production of squalor and the subsequent embarrassment did not curtail the program's use. Because public housing so effectively reinforced the concentration of low-waged laborers and walled off dangerous and property-value-threatening people, perceived necessities by these growth elites, its use continued. Never had such an unpopular but functionally efficient program gone so far. By 1965, all 50 states had public housing with the program sheltering more than 2 million people (Friedman 1980). By 1970, the program housed nearly 1 percent of the nation's population, and every city over 250,000 except one had public housing in place (Solomon 1974). In 1975, New York City operated 116,000 units, Philadelphia 22,900, Chicago 38,600, Baltimore 16,200, and Atlanta 24,700 (Jakle and Wilson 1992). In the 1980s, public housing constituted 15 percent of the total housing stock in Atlanta, 10 percent in Baltimore, and 9 percent in Philadelphia (Jakle and Wilson 1992).

Urban renewal also emerged as a nationwide program that helped sculpt both the content of these black labor-pockets and the economic viability of other housing submarkets. The program, begun in the early 1930s, was re-asserted in the 1950s as a bold way to renew city economies, in particular, to re-make downtowns as economic engines. This often involved destroying "blighted" neighborhoods and downtown blocks to make way for new industrial campuses and office buildings. To a disproportionate degree, these neighborhoods were low- and moderate-income black, with residents systematically placed or steered to other black ghetto spaces (Tabb 1974). These communities tended to be easy targets, lacking clout at city hall. Between 1955 and 1970, over $3 billion in federal aid was earmarked for this program (Weicher 1970). It would, to Philadelphia Renewal Director William Rafsky (1978), "demolish the cancer of blight and clear land for new office and business investment."

These projects were stunning in their obliteration of African-American neighborhoods. Blocks were destroyed and residents forced to relocate to burgeoning ghettos. A sense of relentless modernism infused these schemes, out with the old and obsolete and in with the purported new and efficient. In the process, black ghettos grew in population and areal extent. Fueled by the dual processes of continued in-migration and massive capture of renewal-displaced families, Harlem in New York and Hough in Cleveland increased their populations more than 60 percent between 1940 and 1960 (Osofsky 1967; Hough Neighborhood History 2003). Chicago's Wentworth and Philadelphia's Allegheny West experienced a greater than 50 percent population increase (Wilson 1983). Detroit's sprawling inner city increased its population from 190,000 to more than 570,000 between 1940 and 1960 (U.S. Census Bureau).

By the mid 1960s, urban renewal had proven overly destructive but most importantly, economically ineffectual (Wilson 2005). The program annihilated many more homes than they had built and had displaced more people and activities than they relocated. In Pittsburgh, urban renewal as the city's dominant post-war policy tool razed more than 3,700 buildings, relocated more than 1,800 businesses, and uprooted more than 5,000 families (Teaford 1990). Housing and buildings were typically replaced by utilitarian-style shopping complexes and office towers. In St. Louis, the shockingly slow Mill Creek Valley Project embarrassed city officials and revealed the difficulties of redeveloping severely disinvested land. Mill Creek was called by the *New York Times* (in Teaford 1990) "the questionable spectacle of one of the country's most unsuccessful redevelopment programs." Buffalo's Ellicott project, the model of delay and inaction, began in 1954 and by late 1964 contained only six new single-family homes. The project displaced 2,200 black families in a 161 acre area and proved unattractive for middle-income re-occupation; few at this time wanted to live near these ghettos.

THINGS GET WORSE: 1965–1980

The 1960s and 1970s were tumultuous years for rust belt cities and their black ghettos. Initially, race riots from a sustained black poverty punctuated a period of general societal prosperity battering these spaces. Much mainstream policy and thought, peripheralizing the signs of problematic structural inequalities in inner cities, instead often interpreted this as something less serious – a brief flirtation of poor minorities with anger and discordance (see Kelley 1997). Then, a city and societally punishing 1970s recession deepened deprivation in these ghettos (Harvey 1981, 1985). As Keynesianism and job bases further diminished across America, so too did they in these black ghettos. Thus, paralleling city deindustrialization, shrinking city tax revenues, and increasingly scarred physical infrastructures, was the loss of job opportunities, public resources, and hope among many ghetto residents. Across this time frame, then, initial improvements in the black condition were offset by an erosion of economic and social circumstances (Hacker 1992; Massey and Denton 1994).

The riots of the 1960s occurred in a period of general rising affluence, which was visibly reflected in urban and suburban landscapes and patterns of consumption (see Jakle and Wilson 1992). Rates of automobile ownership, housing homeownership, and income and purchasing power in general society all increased in this period by more than 25 percent (Heilbroner 1976). However, once again national prosperity proved spatially and demographically uneven: anger in these ghettos from persistent poverty, staggeringly low wages, and searing social stigma was pronounced and finally surfaced (Tabb 1974; Teaford 1990). As Swyngedouw et al. (2002) puts it, such cities are brooding places of imagination, creativity, innovation, and the ever new and different. However, they also hide in their underbelly perverse and pervasive processes of social exclusion and marginalization and are rife with struggle, conflict, and often outright despair in the midst of tremendous abundance and pleasure. Seen this way, the riots were anything but surprising.

The largest riot, Detroit in 1967, reported on in stark terms from coast to coast, saw over 800 buildings pillaged and burned (Beauregard 1993). Property damage was estimated at over $70 million (Glazer 1970). The second worse disturbance in the rust belt, a six day riot in Cleveland's East Side (in and around Hough), decimated the area. Most buildings along a 20 block stretch were destroyed or looted (Gillespie 1966). In Newark, four days of riots resulted in 24 people dead and a 15 block area firebombed and annihilated (Winters 1979). Afterwards, a host of Realtors and developers publicly pronounced that they would never do business or set foot in the riot-torn zone again (see Winters 1979). Across rust belt cities, more than 20 people were killed, 2,000 injured, 4,000 arrested, and thousands of buildings partially or totally destroyed (Gillespie 1966). In the aftermath, already trimmed institutional supports declined even more (i.e. banks lending mortgages and home improvement loans, police safeguarding streets, developers investing capital).

It is now widely chronicled that the riots brought a complex mix of hope, despair, new political attention, and new political neglect to these ghettos. On the negative side, physically devastated neighborhoods did not soon or easily recover. Burned out buildings and storefronts in Chicago, Cleveland,

and Detroit frequently lay in ruins for years, and when finally bulldozed, scarred neighborhoods as long-term empty lots (see Vergara 1994). In Chicago's Near West Side, one site of riots, long-term resident Devira Beverly (1991) reflects that the proliferation of empty lots propelled some residents to push for the creation of large farms. More significantly, the riots often took an emotional toll on residents, particularly the elderly, who came to view more cynically possibilities for upgrading their neighborhoods (see Masotti 1968). The sight of local young people pillaging their own stores and homes, as Winters (1979) recounts, was devastating to them.

On the upside, the riots brought attention to the neglect of these areas. A then influential fraction of local and national policy analysts, democratic liberals, identified the core poverty in these neighborhoods and its potentially explosive nature (see Beauregard 1993). The stepped-up institutional fix of expanding social welfare, in the heart of the Keynesian social regulatory project (see Peck 2001), was envisioned. The stepped-up War on Poverty, initiated in President Johnson's declaration of the Great Society in 1964, followed (Judd 1979). The first steps toward a national anti-poverty effort, taken in 1963 by the President's Council of Economic Advisors, became a full-fledged policy initiative with new programs begun under the 1960s Economic Opportunity Act. The Jobs Corps, Head Start, Community Action, Volunteers In Service to America (VISTA), and other programs became a reality fueled by a first-year spending allocation of $1 billion (Judd 1979). The poor, now, were to be both program recipients and key decision-makers in programs. They were to sit on governing and policy advisory boards and be selected by democratic means.

Social spending on black ghettos increased. The 1965-created Medicare and Medicaid programs had their funding increased two-fold (Houghton Mifflin 1991). Urban renewal, widely discussed as a "ghetto-enhancing tool" (amid all of its controversy), had its funding roughly tripled (Wilson 2005), and funding for the food stamp program more than doubled (Levitan 1985). At the same time, 1970 amendments to social security greatly broadened coverage, increased the value of benefits, and indexed them against future inflation (Houghton Mifflin 1991). The result was a deepened and broadened welfare safety net that enhanced the poor's quality of life, and reduced social unrest and enhanced the black population's stake in existing social and economic arrangements. The 1960s riots seared the public psyche for its potential to inflict damage across cities and beyond (see Lieske 1978): this was not to happen again.

However, this "attack" on poverty and ghetto conditions was short lived. These programs, nowhere set as inviolable policy, were soon attacked with the dismantling of Keynesian social welfarism beginning in the early 1970s. An ominous development presaged this: the 1968 election. Republicans claimed 62 percent of the governorships and installed an anti-urban president (Richard Nixon) pledged to decentralize domestic policy and programs. With the riots ended, the black population's needs were peripheralized. Nixon talked about scaling back the 1960s programs and implementing a block grant programatic orientation that would eliminate established "categorical" programs (i.e. federally based aid tied to precise categories with federal oversight). His successor, Gerald Ford, closed the deal, with the 1974 Community Development Act replacing seven major categorical grant programs with a single broad-based initiative. The result was the beginning of a new city–federal relation: block grants, which were to accelerate throughout the remainder of the century and beyond.

But shortly thereafter a new ominous development in cities and society emerged to exacerbate decline in these rust belt cities and their black ghettos: structural changes in local, regional, and national economies. The latest round of something now familiar in capitalist America was at work: the endless quest for better spaces of profitable investment, harnessing new technologies to serve capital's drive for profit (e.g. high speed fiber optics, new computer capabilities), and strategic shifts in investment to new economic sectors (i.e. services). The impact was dramatic and traumatic, leading to a new development: a structural rollback in both the supply of decent city-based jobs for working people and the regime of social entitlements. This shift, well documented as an outgrowth in a shift from "Fordist–Keynesianism" to a "post-Fordist" regime of accumulation, served up three rust belt strangulating trends: the shift from commodity production to service-producing industries, a labor market split into the extremes of low- and high-wage sectors, and the closure or out-migration of manufacturers from cities.

First, the shift to service jobs meant a proliferation of low-wage occupations (the "McDonaldization" of the economy) that paid paltry sums and offered scant opportunities for upward mobility. Jobs that paid below 125 percent of the poverty level nationally rose from 36 percent of jobs in 1973 to 41 percent in 1993 (Marable 1997). This trend has continued: low-wage service-sector employment nationally grew from approximately 24 million people (less than 26 percent of total employment in 1980) to almost 46 million people (34 percent in 2000) (see Defilippis 2004). Second, the labor market split meant that growing numbers of workers had to secure either well-paying, "higher-educational" jobs or low-paying, relatively unskilled jobs. The supply of decent paying, "middle-level jobs" had dramatically shrunk: the city era of heavy industry was over. Third, the flight of manufacturers from inner cities meant a decline in the availability of decent-paying, manufacturing jobs. Chicago, Cleveland, and Milwaukee, for example, each lost more than 120,000 factory jobs between 1960 and 1980 (see Fisher 2004).

This economic transformation of regions and cities, then, had roots in how production was now being spatially organized, carried out through new sets of institutions, and the rise of the service sector (i.e. the shift from Fordism to post-Fordism). It took the form of a shrinkage of and replacement of spatially fixed manufacturing by more mobile, multi-site production, and information-based businesses and services (see Dicken and Lloyd 1990; Warf and Holly 1997). Its visible form was the eclipse of the one-site, assembly-line production of commodities in favor of the more spatially expansive, service-oriented producer. Industrial plants and rust waned, supplanted by services and the appearance of glitz. The result was a boom in production services – computer chips, computers, fiber optics, business services, telecommunication products, health care products, software, fast food, legal services – that would intensify as economic staples in these cities.

For many black ghetto residents in the rust belt, the economic change was disastrous. Many lacked the rarified talents to succeed in this economy's upper end. Forced into the new economy's "lower end," prospects were marginal. Fast-food establishments, retail chain outlets, and grocery stores paid nominal wages and typically failed to provide health benefits. The explosion of Burger Kings, Kentucky Fried Chickens, McDonalds, and Hardees across urban terrains did not mean an explosion in the wealth of their workers. The result of this post-industrialism, to economist Michael Rothschild, was that approximately 20 percent of U.S. workers were significantly marginalized (i.e. made unemployed or suffered from substantially lower wages than before), most of them low- or moderate-income (in Fingleton 1999). Not surprisingly, employment among young black men between 1955 and 1984 fell dramatically, declining for ages 16–17 by 28.3 percent, ages 18–19 by 32 percent, ages 20–24 by 20.3 percent, and ages 25–34 by 11.3 percent (Wilson 1987).

THE REAGAN 1980s

Ronald Reagan's rhetoric and policies further damaged these black ghettos in the 1980s. While this is an unsurprising assertion to anyone remembering these years, the depth of the effect is still profound and shocking. Most directly, his talk and actions unleashed a rhetorical assault on these areas and reduced resource flow resulting in increased poverty, homelessness, and hopelessness. Equally important, his talk and actions were expediently embraced and activated by city growth machines across the rust belt who saw opportunities to fire-up the new potentially lucrative economic engine – incipient gentrification and downtown transformation. In this context, his anti-poor and minority rhetoric had another important repercussion: it set the stage for the 1990s and beyond global trope and the third-wave of ghetto marginalization to seamlessly squeeze these spaces. Thus Reagan's vitriolic rhetoric established a constellation of signifiers about poor African Americans and their neighborhoods that would not go away and was ripe for being built on. In the final analysis, Reagan's 1980s discursive and material re-shaping of these ghettos drastically affected these areas way beyond his presidency.

The rise of Reagan is best understood by considering late 1970s conditions in America. Amidst a slowed world economy circa 1972, America's economy stalled (Martin 1994). An intense mid 1970s recession, weak recovery, and the dramatic economic rise of Japan and China followed, seeming to signal a nation in transition (Eric Swynegdouw's (1992) withering of the postwar "golden age"). The post-war capitalist space economy was moving swiftly into a new phase, post-Fordist flexible accumulation, and America's cities and regions were battered. As many have documented, common thought was seared by the sense of weakened regions, declining cities, faltering international standing, choking inflation, and general economic malaise (see Steinfels 1979; Diamond 1995). As Diamond (1995) notes, it was a time of worry and concern for the state of society, the nation, one's future. Economics, politics, and material well-being seemed unstable, America seemed to be confronting a new world.

In the tussle of competitive politics to explain and respond to America's new predicament, it was no contest: conservative Republicans responded vigorously and effectively. A calculated rhetoric deftly navigated all areas: declining profits in key Fordist sectors, the intensification of international competition, deepening deindustrialization, and growing mass unemployment. Playing to nativist mythology and imperialist sensibilities, Reagan, first, displaced common fears and anxieties to other already emotively charged realms: crime, immigration, public education, and public safety. Second, he offered simple enemies and answers to these problems. A stream of easily identifiable and decipherable enemies, domestic and international, featured "black youth gangbangers," "Hispanic families," "Asian entrepreneurs," and "welfare mothers." These villains, rooted in commonsense understandings (i.e. a history of villainizing) and placed in emotion-laden, simplistic story lines, resonated in this troubled period. Reagan, in the same utterances, brilliantly deployed the rhetorical strategies of "the fear economy" and the never-ending "imperial economy."

Offerings of the imperial economy were arguably at the heart of the whole discursive enterprise. Reagan offered elaborate celebratory rhetoric about American economic and cultural superiority. To Reagan, Americans were far too self-critical and dwelled in needless self-doubt. In Reagan narratives, little had really changed: the U.S. was still the leader of the free world and that would continue. Declarations insisted that "it's morning in America" and "what we once were we would continue to be." Notions of a waning country, he said, were a mythology. Suggestions of a national decline, he said, flowed from liberals, welfare-state politicians, democrats, and disillusioned leftists who inhabited a culture of cynicism and nihilism. Public memories of America's greatness, to Reagan, did not lie; the American people had to renew their faith in this. It was no accident America was still the world's most feared and envied nation, Reagan said, and things were only going to get better.

Using the fear economy strategy to service this aggressive rhetorical project ultimately parceled out much hostility in relation to these rust belt black ghettos. In his feel-good oratory, he also talked much about a problem urban poor that had three parts: cheats who abuse taxpayers, addicted welfare recipients trapped in dependency on the system, and truly needy, disabled people (see Reagan 1982). Only the third type, a supposed small group of the poor, was the sole group society should help. The other two, dodgy poverty recipients, sought to avoid honest labor. To Reagan (1982), "virtually every American who shops in a supermarket is aware of the daily abuses that take place in the food stamp program, which has grown 16,000 percent in the last 16 years." Muscular in mind and body, this fraudulent poor used the politics of racism and discrimination to justify aggressive plunder. The truth, to Reagan, had to be spoken: this group had their moral fiber destroyed by something America had to confront: assaulting welfare policies.

In this process, Reagan spearheaded a public assault against "welfare people" that intensified negative perceptions of the black poor. His narrations commonly featured politically charged terms like "Welfare Queens," "Welfare Kings," "the malignancy of welfarism," and "the pretending disadvantaged" (see Weiler and Pearce 1992). On black poverty, Reagan (in Wilson 2005) said: "The American dream is denied to no one, each individual has the right to fly as high as his strength and ability will take him." Welfarist propagators had to be stopped: America was at war with them as well as with relativist lifestyles, hedonist values, alternative sexualities, atheists and agnostic sensibilities. This mobile army of cheats colonized public spaces, downgraded neighborhoods, created surges in non-marital births, populated local homeless shelters, aggressively panhandled on the streets, and eroded public schools and facilities.

The offering of the ghastly and defiant Welfare Queen headed this litany of resonant icons. All of Reagan's negative notions about black poverty were skillfully collapsed into this one body as a luminous container of values, attitudes, looks, and demeanor. Deftly dressed, color-coded, and behavioralized, she had disdain for mainstream norms, had endless kids and welfare boyfriends, and reveled in hustle and plunder. All that the black ghetto and its population was, she embodied with relish. This decisive and coercive woman proudly wielded food stamps and welfare checks as a marker of her identity and scorned a system that she keenly manipulated. Her place of occupancy, "black inner cities," was saturated with pathology that infiltrated her while she embellished this. In a community of cheaters, chiselers, and culturally bankrupt individuals, she headed the pack. In the images of Reagan, a concerned world abhorrently peered at this spectacle of dysfunction that needed to be changed.

With this rhetoric, Reagan effectively demonized the poor and substantially shrunk poverty programs. The effects on black ghettos, materially and symbolically, were pronounced. His legislative centerpiece, the Omnibus Budget Reconciliation Act (OBRA) of 1981, removed from eligibility approximately 50 percent of the 450,000 to 500,000 Aid For Dependent Children recipients (Levitan 1985). These 250,000 people, disproportionately black and living in inner cities, were left to fend for themselves in an increasingly harsh, service-dominated economy. Another 40 percent of working AFDC recipients lost sizable portions of their benefits (Levitan 1985). OBRA also authorized states to operate work relief programs, what Reagan called workfare. Levitan (1985) estimates that OBRA changes in AFDC pushed

600,000 people below the poverty line in 1982 alone. To planner David Carley (1987), Reagan's policies thrust more than two million people below the poverty level between 1980 and 1988.

In this context, Reagan objected to government involvement in black ghettos. There was a new stress on urban policy which communicated the expendability of "welfarist government" and "pampered city people", rather than the expendability of cities themselves. Reagan was astute enough to realize that the public cared about cities and that urban economies could not be left to die. Cities were thus one of the "national treasures" he frequently spoke of. But he persuasively argued that cities and their poor people could be best helped by turning their fortunes over to private-sector forces. His lack of "traditional" urban programs, he proclaimed, would be ideal. His sole new urban initiative, the urban enterprise zone, conformed to this rhetoric of nurturing the private sector (Burnier and Descuter 1992). It provided tax benefits and regulation-reduced business climates to businesses to operate in these zones. In the end, Reagan and Congress slashed the Housing and Urban Development Budget from $36 million in 1980 to $18 million in 1987 (Burnier and Descuter 1992).

THE POST-1990 GLOBAL OBSESSION

Reagan's rhetoric and policies, sharp and resonant, continued to haunt black ghettos through the 1990s. In a dominant impact, the extending and deepening of these understandings helped shape and legitimate the ghetto-devastating "global trope" and the third-wave of black ghetto marginalization across rust belt cities. The conversion of these devastated spaces to "global ghettos" was spurred by the post-1990 near-hysterical global invoking. Globalization, the new 1990s policy buzzword that seared the public psyche, was seized upon and used as a disciplining rhetoric across rust belt cities that identified a new perplexing economic reality and a "proper" politics of growth, development, and social management. Black ghettos, in this trope, were perpetuated as the resistant city demon, something profoundly anti-civic, which made it easy to steer public concerns and resources elsewhere.

But an important caveat: globalization was not entirely a fiction. Much evidence suggests that the U.S. economy and many city economies in the rust belt became more mobile and far-reaching in the 1990s. Large capital across America progressively restructured itself throughout the late 1970s and 1980s to get to this point (Brenner and Theodore 2002). A punishing six year recession in the early and mid 1970s led this capital to re-examine and re-do its market strategies and organizational structure. By 1990, many corporations had consciously downsized or eliminated inefficient plants, spurned high labor-cost communities, participated in key acquisitions of firms, and aggressively incorporated new technologies (Knox 1997). The result was larger, more powerful entities that tended to engage in more far-flung operations across regions and countries. By 1990, 40 percent of all world trade was between different branches and companies of the same transnational conglomerate (Knox 1997). Ford's economy in 1990, for example, was larger than Saudi Arabia's and Norway's, and annual sales of Philip Morris exceeded the gross domestic product of New Zealand.

In addition, middle-sized capital in rust belt cities often became more global after 1990. They experienced the same 1970s hardships that forced them to examine and re-do market strategies and organizational structure. In Chicago, Exelon and Motorola (collectively with over 11,000 workers locally) grew more than 35 percent in assets between 1985 and 1995 while expanding their network of factories across the globe from 50 to more than 140. Today, Exelon and Motorola concentrate production across America, Europe, and Asia. In Milwaukee, Briggs and Stratton and Harley Davidson reduced their local presences dramatically but increased their assets by 25 percent and 31 percent, respectively, between 1985 and 1995. Currently, each out-sources more than 50 percent of their production to Mexico and Asia. In Cleveland, economic anchors TRW and Eaton experienced a 42 percent and 47 percent growth in budget, respectively, between 1985 and 1995. In 1980, TRW had more than 200 facilities in 23 countries and employed more people in Europe (32,200) than in North America (22,500).

But proclamations in cities about a new reality of punishing globalization can be more rhetoric than real. Simply put, globalization did not uniformly affect all rust belt cities and all economic sectors within them. These cities, post-1990, had different economic bases and ties to global corporations and investment, which made globalization's influence highly uneven. And even in the most "global" of cities, economic bases have tended to be heavily place-dependent. Chicago, for example, has a more "footloose–global" economic base than Indianapolis, Kansas City, or St. Louis (i.e. it contains a significant presence of well capitalized, place-resilient producers like Sara Lee, Boeing, and SBC). These companies have the distinctive combination – substantial assets, far-flung operations, minimal dependence on local conditions to produce – that make out-migration an option. Yet, this kind of producer in Chicago is still the exception rather than the rule, and accounts for a surprisingly low level of total employment: planner M. Fried (2004) estimates this at 18 to 20 percent.

But this reality did not deter Daley and the Chicago growth machine. They talked about the dire need to build a physical and social infrastructure rooted in the supposed necessity to retain and attract "global" producers and investment. In the language of Daley and the growth machine, these producers and investment were the entire economic base. For city economic survivability, key propulsive elements had to be put in place: upscaled historic districts, gentrified neighborhoods, exotic ethnic spaces, and a glistening downtown. Daley was not alone. This drive and language also characterized the administrations of Goldsmith and Peterson in Indianapolis, Archer in Detroit, Slay in St. Louis, Giuliani and Bloomberg in New York, and White and Campbell in Cleveland despite the even smaller presences of footloose producers in these cities (e.g. firms not tied to local labor, raw materials, or local physical infrastructure) (see Cox and Mair 1987).

Thus, the post-1990 global rhetoric echoed loudly and ominously across the rust belt. To be sure, a key impetus was resonant national pronouncement of this. For example, billionaire financial speculator and media guru George Soros (1998), the media's poster boy for globalization, talked frequently in public forums and T.V. studios about this, comparing it to the dawn of the machine age and the age of reason. To Soros, a new hyper-competition for jobs and investment faced America given their footlooseness. Producers, to Soros, could now seamlessly move in and out of regions, countries and continents. Split-second decisions conveyed by frenetic telecommunications systems could instantaneously move jobs and investment and coordinate far-flung operations. To Soros, the world had irreversibly changed: policy had to be responsive and make the country and its cities more competitive.

The national media unceasingly promulgated the supposed new reality of globalization in the 1990s. For example, in *Business Week* mentions of "globalization" or "global economy" escalated from 160 in 1990 to over 290 mentions in 2000 (Miller 2003). Similarly, a content analysis of forty newspapers and magazines found 158, 2,035, and 17,638 stories using the term "globalization" in 1991, 1995, and 2000, respectively (Miller 2003). To the trusted and authoritative *New York Times* (in Foster 2002), the reality of globalization was undeniable: "[it] has set in as a . . . fluid, infinitely expanding and highly organized system that encompasses the world's entire population, but which lacks any privileged position or 'places of power'." To national commentator and educator Rik Anderson, "the contemporary world is characterized by a historically unprecedented international or global character." It is, to Anderson, "a long-term historical trend toward a . . . globalized human condition" that we can't be removed from.

Local media and politicians across rust belt cities also forcefully unleashed this rhetoric. To U.S. Conference of Mayors representative Marc Morial (2002), "every mayor [now] needs to be a player on the global economic scene. Every mayor must recognize that our cities, no matter how big or small, are important to the new global economy." Morial "challenge[s] every single mayor to spread th[is] message throughout your city, to your editorial boards, to your Chambers of Commerce, to the pulpits of your places of worship . . ." Similarly, a communiqué issued by the Transatlantic Mayors Summit in 2000 said: "mayors . . . are no longer the traditional city planners of the past 100 years, but rather ombudsmen and innovative leaders [that] . . . conduct their own foreign policy." "For the good of their citizens," continues the communiqué, "mayors will increasingly enter the international arena and become global players."

This rhetoric, of course, partially reflected the sense that cities faced changing times. A pervasive sense of impinging globalism was in the local air (see Thrift 1995). But this rhetoric was also recognized as a kind of expedient resource that could assist a central desire of city elites: to fully shift local politics to a concern with resource attraction and forge land- and property-vibrant cities. This invoking could thus fully transition policy to a focus on producing wealth and intensifying land values that had begun in the 1980s. This goal proceeded from the idea of cultivating enhanced city prestige, creating wondrous opportunities for conspicuous consumption, and building upscale housing complexes that could swallow up "under-utilized" land in their wake. This thrust promised something intensely desired by local government, with shrinking federal aid throughout the 1990s: more tax revenues. For local developers, builders, Realtors, and speculators, it could facilitate tremendous profits in local land and housing markets.

While multiple voices espoused this new ominous global reality, local governments proved remarkably audacious, and a discussion of them is important. Local governments, a crucial actor in these growth machines, realized that this political transition, with them out front as policy providers, would service their major constituency – business – which would also help them. Local government, with the recent ebb and flow of economic conditions and political swings, had too much at stake not to support capital (see Dear and Clark 1984). As Bob Jessop (1990) notes, the local state takes on many appearances and forms, but is compelled to bolster capital, fundamentally because capital striking out on behalf of itself simultaneously advances government goals and designs. Put another way, local government, without capital, is a kind of warrior without a sword, an ambitious apparatus without the mechanism to realize its designs. Far from being a passive and simple instrumental supporter of capital, then, local government enacted this support as an active and goal oriented apparatus.

A key resource enabled government to justify its intensified support for capital via this policy shift: historicity and its stock of understandings. Government, as discussed in Terry Eagleton (1991), has been enabled historically by the public's general identification of business as the motor of urban vibrancy. The roots of this are complex, and appear to lie in America's distinctive privileging of a private-market ethos (see Jakle and Wilson 1992), but also in diverse discourses through the twentieth century – on city growth, urban redevelopment, civic health – which have relentlessly cast this group as the engine for city solvency (see Beauregard 1993). Notions of business have fluctuated across the century, but have been variously identified as missionaries of the public good, skilled technicians to convert semi-chaotic landscapes into taut economic instruments, ideal cultural leaders to reveal proper civic tastes and sensibilities, and meritorious social–cultural role models (Boyer 1983). The social and spatial of cities, in short, have been said to have internal laws of ideal functional organization and development with business elites most able to progressively engage these.

At the same time, this exuberant voicing of the new ominous global reality was a practical stroke in another way. Simply put, administrations that pushed this new politics could enhance opportunities to gain public support and get re-elected or re-appointed. To understand this, we must realize that urban populations and America turned more conservative after 1980, and support for America's "Golden Boy" (business) was more than ever sound politics. Thus, the local state (and other growth machine operatives) did not merely make social and political realities, they also responded to them. While fears and anxieties about immigrants, minorities, and the poor, re-awakened in the Reagan years, did not go away, the private market was continuously extolled as the solution to social, economic, cultural, and political ills. A startlingly conservative national media that formed in the Reagan years, featuring old and new pundits like Mona Charen, John McLoughlin, Mike Novak, Brit Hume, Tony Snow, Walter Williams, Thomas Sowell and Emmett Tyrell, obsessively held up the private market as the beacon of hope to move America forward (see Wilson 2005).

In this sense, political viability post-1990 was enhanced by demonstrating political toughness and entrepreneurial sensibilities rather than being "welfarist" and "redistributional" (Beckett 1997). Times changed, to Beckett, and the rhetoric and institutions Reagan fostered came of age and deepened. In this context, many mayors saw their political futures tied up with key processes: upgrading city prestige, building glistening downtowns, attracting celebrity companies, and stabilizing tax bases. Eradicating poverty in black ghettos was not seen as the ticket to political success; demonstrating bold, upscale restructuring was. The litany of brash pro-neoliberal mayors in rust belt cities post-1990, whose solution to all kinds of city issues (fiscal, social, political) was the private market, reflects this: Giuliani in New York, Norquist in Milwaukee, Daley in Chicago, Goldsmith and Peterson in Indianapolis, Slay in St. Louis, and Campbell in Cleveland.

FORMATION OF THE GLOCAL GHETTO

This global rhetoric as a complex, constructed vision spurred the post-1990 third-wave of black ghetto marginalization that sculpted the glocal ghetto. This rhetoric was not the only political force that spurred this, but was the prominent one (see Wacquant (2002) on these other forces). First and foremost, this rhetorical formation was built on previous waves of restructuring and their rhetorical formations, which melded past and present processes. As in Doreen Massey's (1999) geologic metaphor, a past history of sedimented rounds of restructuring and guiding discourses built a distinctive space that set it up for this latest round of transformation. The post-1990 glocal ghetto thus formed in rust belt cities not in a dramatic emergence but in a continuous and paced neglect and decline as the collective weight of past rhetoric, resource flows, plans, and policies collided with its corollaries in the present. The past, ultimately, primed these spaces as objects to be seized and re-made in the present.

In this context, the impact of the post-1990 global hysteria on black ghettos was pronounced. Most immediately, government resources and supports were steered away from these areas with potent consequences. While this process included numerous government programs – tax abatements, tax increment financing, human and social service funds – especially significant was the re-orienting of community development block grants (CDBG). This program, the major government instrument to help urban neighborhoods across America since 1974, dried up in these poor neighborhoods. An already diminished amount of block grant funds allocated to these cities (a program still characterized by a flexible usage that could span housing upgrading, economic development, and job creation) were increasingly used to upscale downtowns and beyond. As a New York City planner supportive of this resource diverting commented to us, "times had changed, and block grants as flexible funds could be re-spent . . . as a general city improvement pool of funds . . . now brave voices were required in our downtowns . . . fighting for . . . far-sighted revival that we needed . . ."

At the same time, provision of tax abatements in these rust belt cities re-surfaced with a vengeance in the 1990s to further reduce government funding to these poor black neighborhoods. Tax abatement schemes mushroomed in the early 1970s as growth coalitions scurried to find new strategies to generate revenues with the devastating national recession of 1973 (Wilson and Wouters 2004). These schemes again heated up with the early 1990s attempt to mold the stepped-up entrepreneurial city (Mokhiber and Weissman 2003). Businesses, corporations, and real-estate interests gained but at the expense of depleting a central funding source for the creation of housing, jobs, drug-treatment programs and facilities, after-school programs, and the like in these ghettos. In the process, of tax abatements evolved as a kind of entrenched culture, which made this subsidy expected in the everyday workings of these cities. As a planner in Cleveland told us:

> we dole out tax abatements all the time, it's just common city practice . . . one of our tasks. Does it work? Probably not, maybe a little, but everywhere else, you can get these too. We're just supposed to give these incentives to businesses . . . it's like saying [to a business] hello and welcome to our community.

At its core, this resource diversion from ghettos to downtowns was offered up as a city-survival decision. Its appeal was the imposing of an imagined unity, control, coherence, and expert skills onto the city. But it was also a consciously racialized and spatialized project. Most notably, global prognosticators knew that impoverished black neighborhoods were easiest to squeeze to help re-entrepreneurialize the city (in the era of dwindled federal support). Pulling these resources from ghettos, as ongoing human endeavors, involved a "classing" and "racing" of these neighborhoods as profligate and unworthy. In the extreme, these terrains were demarcated as degenerative and culturally contagious via a core theme, enhancing city competitiveness in new global times, and lesser themes like "cleaning up streets and parks," "suppressing public nuisances," and "enhancing livability." Poor blackness, paraphrasing Wacquant (2002), was again marked as a "principle of vision and division" that revealed for all to see the city's supposed civically unproductive, residential underbelly.

In post-1990 Cleveland, for example, the administrations of Michael White and Jane Campbell drove the city's new uneven development. Each government, populist in thrust, mirrored the reality of neoliberal policy and project: a rhetoric of small government barely disguised a profoundly interventionist local state. Its interventions were fueled by a belief that urban real-estate could be heated up by a mix of physical impositions and spatial banishings. This entity, focusing on the core, engaged and regulated people, land, and institutions by offering drastic welfare cuts, rigorously policing the streets (declaring war against panhandlers and the homeless), extolling and administering Workfare, attacking the public unions, providing more than $50 million after 1990 to "culturalize" downtown, and offering new zoning codes and variances. The succession of governments during this period in traditionally democratic Cleveland were, as Hennepin (2004) notes, the most active in the city's history.

To drive restructuring, Cleveland was repeatedly cast in a cautiously optimistic global frame. Reporting and oratory presented the city as threatened by globalization, but as a historically resilient place, once again had to act ingeniously to survive. The Civic Task Force on International Cleveland (2003), for example, called the city a place [with] "an . . . opportunity to revitalize . . . through continued internationalization of . . . population and employment opportunities . . . economic revitalization [can] occur." Its black ghettos, alternatively, were repeatedly signified as dysfunctional and best marginalized amid new global realities. Thus, strategies to deal with the city's massive Eastside Black ghetto, said popular Mayoral Candidate Robert Triozzi (2005), should involve "bringing back police mini-stations . . . and put[ting] Cleveland back on the economic development map." Cuyahoga County Commissioner Jimmy DiMora (2005) noted a solution: "to build on Cleveland's arts and cultural industry as if our community's future depend[ed] on it." In the name of city progress and evolution, it was communicated, black ghettos could be marginalized for the broader city's good.

In this context, government resources obsessively concentrated downtown with one thing privileged: sellable culture. Cleveland, long the quintessential blue-collar American city had, post-1980, experienced social and physical transformation through a traumatic deindustrialization. This deindustrialization was particularly destructive between 1971 and 1981, with closure of the U.S. Steel plant, General Motors' Coit Road factory, the Westinghouse lightning products plant, and six General Electric plants (Warf and Holly 1997). In the mid 1980s, gentrification was crafted in select downtown neighborhoods and the core was revitalized. The key pieces were the Rock and Roll Hall of Fame, the Old City and the Flats gentrification districts, sparkling Gund Arena, and new hotels and clubs. After 1990, the growth machine built on this transformation and labored to symbolically bury the image of "the Mistake by the Lake" under the hype of the Chamber of Commerce's "Comeback City." But this post-1990 restructuring differed from previous rounds of redevelopment: these projects were not meaningfully job-creating or economically propulsive. They were, most distinctly, enormously lucrative for builders and developers (e.g. Gus Georgalis, K & D Group, Fred and Mark Coffin).

In the new restructuring, a deeper privatopia of wealth and discerning pleasure-taking was etched into the city, whose stability and fortification rested on a social and spatial banishing of heterogeneity. Its centerpiece were these projects – Dear and Flusty's (2001) interdictory spaces – designed and regulated to systematically exclude particularly poor African Americans. A suggested character of problem people, (i.e. their class and cultural position), diverged from the cultivated texture and purpose of these spaces, and were to be repelled. These territorialized pleasure playgrounds were protected through now well-tested procedures: rigorous policing, assiduous zoning, and youth and gang ordinances that problematized if not criminalized styles and bodily appearances. As one city organizer opposed to the new restructuring told us in discussion,

> Cleveland claims it has to upgrade and look appealing to all kinds of investors and speculators . . . in the name of this, the downtown vicinity is re-made in this image [of the re-entrepreneurialized, globally-competitive city] . . . and the rich get richer, the poor got poorer.

Thus, its black ghettos worsened. East Cleveland, predominantly black and low- and moderate-income, was most decimated by the concerted effort to restructure the city. Block grant monies that once helped this area's poor black Hough, Glenville, and Fairfax neighborhoods were substantially diverted to build the new upscale Cleveland. Hough experienced a more than 40 percent decline in funds received from 1990 to 2000, with a dramatic decline in subsidized day care provision, an affordable housing construction program, and a job creation initiative (Skrabec 2004). City Councilor M. Wallach called this withdrawal of resources

> anything but surprising, a response to the new realities that securing funds to help the poor was now extremely difficult . . . the cuts really hurt the Hough area . . . the poor now had to help themselves . . . practice self-uplift . . . start to think entrepreneurially . . . this was the new rhetoric of a supposedly progressive Cleveland.

To make matters worse in these neighborhoods, land and property here (like the rest of the city) was more strenuously subjected to market rule. Along with black ghettos floated in public discourse as culturally failing and productively inept, this was a recipe for neglect and abandonment. Imposing market rule in these impoverished neighborhoods had different repercussions compared to this imposition across the broader city. Market rule disciplined much of Cleveland to revalorize; this same rule deadened land and property worth and its institutional supports in these ghettos. Simply put, private-sector investors did not want to reinvest in housing, jobs, or venture projects here. As planner Chris Jenks put it, "as places for profit . . . these neighborhoods were barely on the map." Thus, Cleveland Planner B. Hennepin's (2004) notion of "the continued black ghetto quandary" and his plea for "all of Cleveland . . . and the ghetto . . . to find resuscitation via the market . . . which will impose a needed order . . ." cast the die for something predictable: the further abandonment of these spaces by the private-sector.

Indianapolis, under mayors Stephen Goldsmith and Bart Peterson, further chiseled into its local fabric the new uneven development. These governments, too, were deceptive and illusory neoliberal projects. Bold oratory of a retrenched government barely concealed a forceful local state that strictly regulated and managed land, property, the employment base, and social service provision. Like Cleveland, the goal was to heat-up urban real-estate via a mix of physical impositions and spatial banishings. Its post-1990 arsenal offered, among other things, new, stepped-up policing methods (Project Saturation, the Zero Tolerance Team), new land-use control devices (e.g. removing the homeless and panhandlers from downtown streets), Workfare and No Child Left Behind, and the deeper subsidizing of downtown redevelopment (see Grunwald 1998). A conservative local state, staunchly Republican, keenly inserted itself into local lives and spatial configurations.

In the process, public resources were massively diverted to build an upscale downtown. Like Cleveland, these projects were enormously lucrative to growth machine members (e.g. the Simons, J. Scott Keller) and were not meaningfully job-creating or economically propulsive (see Wilson 1996). The rise of Circle Center Mall, Conseco Fieldhouse, White River Park, and new posh hotels were its centerpiece. Gentrification also intensified in its near downtown neighborhoods anchored in two spaces: Lockerbie Square and old North Side. By the end of his second term (1992–99), Mayor Steven Goldsmith had presided in funneling more than $1.5 billion in new downtown spending (IndyStar.com 2001). Sports stadia, upscale housing, restaurant rows, and theater blocks replaced acres of working-class neighborhoods and open spaces. "Indiana-no-place," in short order, gained national notoriety to become the Republican-hyped model for ideal city redevelopment in America (see Goldsmith 2003).

But in the process, funds to meet the poor's housing and social needs were cut and often superficially used. Most notably, block grants for Eastside neighborhoods declined by more than 40 percent while the city-wide decline was approximately 25 percent between 1990 and 2000 (City of Indianapolis 2003). The 35 once subsidized day-care facilities and eight counseling-drug treatment centers dwindled to eight and four by 2000 (City of Indianapolis 2003). To make matters worse, neoliberal forms of intervention came to dominate in these areas. Funds to "distressed neighborhoods" went mainly to two sources: Community Development Corporations (CDCs) and the National Center For Neighborhood Enterprise (NCNE). The city's seven major CDCs, substantially controlled by the conservative City County Council, used funds to mainly paint homes, fix up houses, repair torn streets, and enroll residents in entrepreneurial programs (Maher 2003). They, like other CDCs across Urban America, had become remarkably corporate and deradicalized by the 1990s (see Defilippis 2004). CDCs, operating in ascendant neoliberal times, often functioned out of necessity as corporate collaborating, embracing entities (see Porter 1997).

The other source of funds for distressed neighborhoods in Indianapolis, NCNE, is conservative mogul Robert Woodson's national outreach center. His city development model re-entrepreneurializes social climates and physical spaces by nurturing individual responsibility, business acumen, and supplanting "bad" culture for "good" culture. Its increased use across inner city America, initially in trials, appeared widely as central government policy across the rust belt in the 1990s (it was embraced and widely used by John Norquist in Milwaukee, Rudolph Giuliani in New York, and Stephen Goldsmith in Indianapolis). Its implementation in Indianapolis emphasized one pillar of this, faith-based programs and church led social expertise to revitalize neighborhoods. At its core, it turned more than 15 churches into major providers for social counseling, job expertise, drug control, and micro-enterprise classes (see National Center For Neighborhood Enterprise 2003). Each operated in the context of proffering faith-based and "good" cultural values: Indy's black ghettos were to be re-shaped as compliant, low-wage labor pockets.

These Indy ghettos ultimately fell prey to a maxim that guided the local growth machine: the need for growth leaders to be brutally efficient and partition the city into separate social spheres in new hyper-global times. Ghetto spaces were ostensibly ill-suited to the exigencies of a transforming world and city order as the full power of technical and functional thinking pushed them to the margins. As one city planner noted, "the new [global] times . . . the new order of the day . . . without blinking an eye, the choice was obvious, we had to re-make the downtown or else . . . the result might be . . . decay and hard times could prevail." The result was to further the creation of not one city – Indianapolis – but a multiplicity of disconnected places. What emerged was an expansive urban terrain simultaneously stable and on the edge, islands of differentiated spaces bonded only by the sense of being in a place called "Indianapolis." Downtown celebrated spaces – the new consumption playgrounds – were forged as places of hard rock stability, but at the expense of purging dissimilar people and spaces in and around it.

Neighborhood associations and housing groups in Indianapolis and Cleveland did not totally wilt in the face of this reality: some operated "below" CDCs but with severely depleted funding. As one head of a housing group in Cleveland told us,

> it's become much more difficult to help the poor . . . the money was no longer there, we got money from the local CDC which is block grant dollars, but priorities of City Hall changed . . . It became in to fix up downtown and its public hotspots, out with the needs of the poor.

A similar story comes from the head of a social service agency in Indianapolis.

> The CDCs complained about a steep cut in their budgets from City Hall, but it's us who were bearing the brunt of the turn away from the poor. It's [helping the poor] just no longer popular and seen as important to city improvement . . . The City more than ever catered to the needs of builders and developers who want to gentrify . . . that's the priority.

Yet it must be added that this public and private turn away from poor black neighborhoods was never complete. These governments and growth machines could never entirely abandon the benevolent-purpose rhetorical formation that had once been central to government and growth machine presentations of their duties prior to the ascendancy of neoliberal times (see Marcuse 1978). Thus, this resource shrinkage to and re-entrepreneurializing of black ghettos was proclaimed as one kind of human resources outreach. In short, this was supposedly what would best help these residents and spaces. At the same time, City governments continued to hold up CDBG as a kind of pre-neoliberal, persistent emblem of a still active government in poor communities: this was a carryover from the heady days of the mid 1970s when Republican politics nationally boldly ushered in CDBG.

In this setting, these growth machines increasingly trumpeted a new, supposed innovative program: Workfare. With passage of the Personal Responsibility and Work Opportunity Reconciliation Act of 1996 (PRWORA), states and cities could constitute their own "welfare-to-work" programs. Within the general federal guideline that welfare recipients had to now perform waged work (for a maximum period of 5 years), rust belt cities quickly offered programs. With much fanfare, these programs were declared the latest and most effective response to poverty and dilapidated neighborhoods. Milwaukee's mayor John Norqueist (1998) proclaimed their program the answer to "the very negative . . . welfare system . . . that treated pathology and so encouraged it, pa[ying] low-income people not to work." Rudolph Giuliani termed Workfare "the most realistic program you can have for a city . . ." (in Online News Hour 1997).

But Workfare fit best a key theme being incessantly articulated in these cities: the global rhetoric and the new need for cities to be responsive. Workfare was resonant as an innovative new outreach to a problem class that promised a long needed disciplining of this group. It would compel welfare recipients to be responsible and civic, thereby helping them, and create economically productive contributors in the new ominous global times. The new global "reality," wielded like a cudgel, purportedly required that places and populations mobilize energies, be productive, and strike out innovatively. Workfare, forwarded as a key policy instrument in this, offered a brute functionality that was tempered by an aesthetic of civic gain and personal social uplift. Workfare, in short, was sold as punishing, but punishing with a purpose. As a workfare administrator in Chicago, B. Lloyd said, "our program is innovative and tied to the new realities of Chicago and beyond. People receiving welfare should be working and have to be working . . . the city needs this, their self esteem needs this."

But Workfare appears to have damaged many of these black ghettos and helped create their latest form, as a wealth of studies shows. For example, on the labor market front, Nichols and Gault (1999) find that Workfare in Milwaukee tends to place participants in the lower reaches of occupational structures and keeps them there. The bulk of jobs are low-wage, dead-end and fail to lead to better paying, stable jobs. ABT Associates (2001), commissioned by the State of Delaware, finds that the state's late 1990s strong economy and non-nonsense Workfare program did little to boost program participants. Few participants achieved economic success and were able to move out of poverty. To the Education Partnership Program (EPP) (2004), Workfare helps little with few jobs out there to secure. Employers want workers with literacy, numeracy, and information technology skills, to EPP, and Workfare jobs largely fail to provide these. Workers end up in dead-end, demeaning jobs that perpetuate poverty.

Workfare has been just as damaging to participants on the housing front. New York's Community Voices Heard (1999) surveyed 500 city residents on Workfare for more than six months and found no improvement in their accommodations. Many lived in substandard housing that remained virtually unchanged despite participation in Workfare. Even the conservative Center On Budget and Policy Priorities (2002), in studies of Workfare in Chicago, Detroit, Cleveland, and Los Angeles, found deep problems. Most participants still experienced severe cost problems in housing, inadequate provision of shelter, and hopelessness and deprivation. Minimal wages from work provided no stimulus for landlords or developers to improve the supply of affordable housing. Many studies, in sum, document Workfare's reducing of the welfare rolls in rust belt cities, a political plus for those in office, but also chronicle its failure to materially upgrade program participants.

But black ghettos were seared in another way by the global hysteria: they were increasingly policed and cordoned off as problem spaces. This global rhetoric, we now know, implored the public to recognize the twin necessities of cultivating new aesthetic areas and protecting their character. The oratory, implicit and explicit, was thus as much about defending these terrains upon construction – gentrified neighborhoods, upscale ethnic spaces, historic districts, aestheticized public spaces – as forming them. The global project in these cities, in this sense, was multifaceted. Milwaukee Mayor John Norquist (in Riverwest Currents 2003), for example, succinctly communicates this in brash oratory: "a bad environment kills community . . . you can't build a city on pity . . . A good one doesn't necessarily foster community . . . [We need] gentrification . . . [and it needs protecting]"

Cities, seen this way, had a common enemy: pervasive black ghettos. The need to isolate them, in communication, was a civic responsibility. This communicating often took the form of depicting these neighborhoods as mobile and troubled civic problems (i.e. they "explode[ed] in population . . . are lined with dilapidated buildings . . . have burgeoning youth populations and gangs . . . are storehouses for spreading poverty and blight . . .)" (c.f. *Chicago Tribune* 2003; Cline 2004). Socially acceptable language, invoking the likes of cancerous gangs, blight, the mean streets, and black lifestyles, played to established caricatures and stereotypes of a racial space. The rhetoric, both controlling and inciting, widely portrayed a distinctive ghetto world that was threatening and vile.

The worth of these neighborhoods was now to be measured in a new way: by their degree of contribution to "the global project" (e.g. attracting investment, nourishing cultured spaces). The black poor, by virtue of their supposed non-contributory status, were to be banished to the margins as objects of concern in this new city reality. The rationale for the banishment was powerful: these ghettos were obstacles to what the city needed to become. The public good was at stake, appeasing ghetto constituencies had to cease. This placing the black poor on the spatial margins was deftly communicated by Indianapolis Planner S. Holding (2004):

> [we now] tackle city problems and embark upon growth aware that the city must be changed and patterned for a new global economy . . . the public knows that we have to adapt to new times that determines our patterns of expenditure and new investments . . . the downtown is the key, the fulcrum to it all . . . all other [city] areas should be considered in this light.

This rhetoric enabled these cities to rationalize the use of diverse tools to isolate their black ghettos: stepped-up policing, use of curfews, increased stigmatizing of black kids, re-doing zoning, and increased surveillance of streets and public spaces. Diverse agents – planners, the police, City Council staff, mayors, and the media – pressed ahead with these undertakings, which usually did not rub excessively against common sensibilities or appear unduly outrageous. These tactics, in short, seized the ground of normalcy and legitimacy. Some tactics were new, like use of curfews and increased street surveillance, and tended to need more careful elaboration to achieve legitimacy. Other tactics, already established, like rigorous policing and youth monitoring, could subtly shade into being more stern and punitive. This mix of acts sometimes generated little public controversy, at other times they were problematized (see chapter 6).

In Chicago, stepped-up policing on the South Side "Black Belt" and Near West Side in the 1990s resulted in a dramatic increase in the number of people arrested for drugs (see for example *Chicago Reporter* 2000). Many appear to have been confronted in the areas proximate to the gentrifying University Village, Pilsen, and Ukrainian Village areas (Balkin 2002). At the core of this was a 1990s curfew ordinance rooted in the rhetoric of containing and controlling gangs (Thomas 2004). Chicago Police Superintendent P. Cline (2004) echoed these thoughts: "street gangs today are much more sophisticated and more violent . . . they [must be contained] . . . the amount of money they take in is staggering." To Deputy Chief of Organized Crime Mike Cronin (2004), "the gang problem" required sterner measures: "[now] we are going at people different . . . we're doing Title 3 wire taps, we're doing body wires, we are doing all sorts of things we never did before. About 40,000 curfew violations in Chicago have been issued yearly, many involving black youths under 17 who were outside after 10:30 weeknights (First Amendment Center 2004).

Chicago's desire to isolate black poor undesirables, a relentless pursuit, even included attempts to shame two frequently "out-of-place ghettoites" – hookers and johns. Thus, a police department web site was expanded and packed with pictures of accused prostitutes and johns arrested for soliciting (what writer Mark Konkel (2005) called the state's widely publicized "anti-escort service"). Put on display were mug shots, names, addresses, and descriptions of these people to shame them into giving up these activities or confining them to peripheral (i.e. their own neighborhood) places. As Mayor Daley (in Konkel 2005) put it, "In Chicago, if you solicit a prostitute you will be arrested and when you are arrested people will know . . . We will place your name and pictures on the police department web site . . . I don't have to tell anyone how fast information travels on the internet." This state-sanctioned threat of humiliation, today, flourishes as a widely accessed web site that thousands download daily (Weinberg 2005).

In New York City, Rudolph Giuliani ("Mr. Global") became mayor in 1994 and presided over a massive re-making and insulating of Midtown Manhattan. The cultivation of upscale restaurants, lavish townhouses, and trendy shops barred "ghetto" and riff-raff influences by land-use management strategy and mobilizing the police (see Smith 1996). Planners, now, were mandated by the city to create "a nub of affluence" that was to be extended and protected (see Smith 1999). Similarly, the police, to Koolhas (2003), served as a cadre of roving harassers to rid the streets of any sign of class "otherness." Sweeps harassed kids, dislocated the homeless from sidewalks, and barred youth from standing still. Midtown Manhattan, it was communicated, was a space of affluence and privilege, and poor blacks were to purge this from their activity spaces or face the consequences. These streets, to Koolhas, became class and race territorialized. For kids from Harlem, South Bronx, and Bedford-Stuyvesant, Midtown Manhattan became a no-go zone.

At the same time, a host of other black "interlopers" into downtown were denied access via regulations or harassment. Bicyclists became the target of a police crackdown, especially messengers and delivery workers (disproportionately black or immigrants), who often served in a semi-legal "gray economy" (see Weinberg 2005). Their bodies and their frequent servicing of a kind of economy were not welcome in Giuliani's new global spaces; the proper place was their own neighborhoods. Squeegee men, also disproportionately black and poor, became one of Giuliani's civic demons when he ran for mayor in 1994 and 1998, promising an uncompromising crackdown on them to improve quality-of-life (Weinberg 2005). In one easily pointed-to body was a race and occupation that was folded into an identity proclaimed inappropriate for the cultivating of the new global midtown. With installation of police video cameras in the likes of Washington Square Park, smaller parks across midtown, and in smoke detectors at City College of New York, the means to perform this surveillance was bolstered.

Cleveland's experience was much the same. In the 1990s, efforts to protect a revitalized downtown from social and economic "blight" involved sweeps of the homeless and African-American youth from the core area and stepped-up policing of nearby black neighborhoods (National Law Center 1994; Wills 2004). In this rhetoric, two prominant rhetorical themes, safe streets and downtown respectability, communicated the goal to make the core affluent, homogeneous, and a space of rigorous surveillance and policing. Safe streets, it was said, are essential, they were the most concrete expression of the social contract. To have disorder in the streets, to one planner we talked to, "[was] to kill the foundation of city social life, the daily human round." Similarly, the notion of downtown respectability sanctioned the vision of a choking core chock full of undesirables, notably black youth, aggressive beggars, hustlers, litter louts, and junkies, that had to be controlled.

Thus, the harassing and arrest of black and Hispanic kids became routine. To critic P. Rustack (2002), "police were everywhere in the downtown and especially near the new hotspots . . . and if you didn't see them, you kind of felt them, it became a real regimented kind of area." Perhaps unsurprisingly, Cleveland was put on the National Law Center's list of the five most brutal sweep cities in the U.S. in 1994 until a successful lawsuit in 2000 made this police harassment illegal (Wills 2004). The goal was fundamental: to erase black and other bodies from an emergent zone of affluent consumption and pleasure-seeking. A resident we talked to, who observed Cleveland's post-1990 crackdown, described a process of "aggressive officers . . . sweeping through the streets . . . bullying and intimidating kids, street people, and especially black youth." Also, he adds, "black neighborhoods have been infested with cops cracking down." "To be black and poor," he said, "[was] the worst."

Even Columbus, Ohio, experienced a heavy dose of 1990s global hysteria, downtown revitalization, and stepped-up attempts to isolate black ghettos. A well chronicled policy of racial profiling and black harassment, central to this, established activity spaces and living spaces that purged poor blacks. Pulling over black drivers outside "their neighborhoods" has been a long-term, persistent complaint in Columbus. But in the 1990s it got so bad that one officer, Tyrone Thomas, finally broke rank and asked an independent review board to monitor his colleagues (see *Columbus Post* 2003). Before the press, Thomas said:

> The city needs to address [the persistence of] racial profiling, excessive force, and deadly force . . . This city needs to address problems in the black community where our police officers are not community policing . . . Community policing is when a kid runs up to the officer, not runs away from them.
>
> (in *Columbus Post* 2003)

Thomas's charges, an independent investigation discovered, were totally accurate. A widespread pattern of racial profiling and harassing blacks was found in routinized police actions. Thomas is currently off the police force but remains one of its harshest critics. But the effects of profiling and harassment were tangible: the police became widely feared in Columbus's black neighborhoods. Many African Americans were afraid to drive outside their communities at the wrong time, believing they could become innocent victims of a harassment policy tied to controlling black activity spaces (see Fitrakis 2000). The Columbus case was unusual in one respect: attempts to isolate black ghettos from downtown upgrading extended to monitoring and managing vehicular traffic: black body encasement was to be total and complete.

3 The global trope

> If you don't run with the Global Herd and live by its rules . . . accept the fact that you are going to have less access to capital, less access to technology, and ultimately a lower standard of living.
>
> (Friedman 1999: 168)

In 1998, Ramesh Diwan issued perhaps the strongest counter to the notion of globalization as a new universal reality that dominated all places. He said:

> globalization has become a buzzword . . . a popular term in the lexicon of bureaucrats, consultants, journalists, and policy analysts; only a few years back it could not be found in a respectable English dictionary. Like other similar buzz words . . . it is rarely defined but used to promote arguments favoring business interests . . . Repeating it ad infinitum they have given this assertion the advantage of familiarity . . . It has acquired . . . a legitimacy . . . such is the power of subtle propaganda.

What has emerged from this, to Diwan, is something profound: "wealth for some . . . [and] growing attendance in soup kitchens, homelessness, and income inequalities" for others. To Diwan, invoking the notion of globalization unleashes an anxiety that today controls and manages many things: people, places, spaces, and institutional actions.

Some like Diwan now recognize "global-speak" as a politically-expedient rhetoric. Yet, somewhat surprisingly, too many of these people (including Diwan) continue to perpetuate a myth about this: that it is a blunt and straightforward assertion. Discussion typically identifies a brutal, simple rhetoric that incites the public to near-instinctually support a politics of resource attraction in cities, towns, and nation. But to believe this is problematic: we fall victim to the very rhetoric that we seek to critically engage and understand. This chapter reveals the complexities of the post-1990 global trope in rust belt cities that has often been dimly conceived by the public and academics. This unearthing is important in the study: it moves the analysis beyond the deceptive affixing of this rhetoric as simple and straightforward. To understand the 1990s rise of the new global ghetto, we must expose this rhetoric as a complexly conceived and active human accomplishment.

This exposition borrows from Norman Fairclaugh's (1992) idea of the coercive discursive formation to understand the global trope's complexities. Such formations, to Fairclaugh, are the basic building blocks of knowledge. They, as impositions into the realm of common thought, map swaths of understanding about issues and events across the terrain of societal understanding. Served up for public consumption are two key elements, visible, thematic-specific issues (e.g. discussions about the truth of globalization, urban crime, city redevelopment) and the less visible underpinning of staged, fabricated realities to support the presentation of these issues (notions of people, places, processes, the world). This unity of theme and discursive base collectively delimits something crucial to an inherently coercive ontological project: accepted ranges for human conceptual possibility. The rhetoric, in the process, strategically annihilates any sense of a discontinuous, ruptured, fabricated, or contingent perspective onto the world.

In this context, two key points about this guide the dissection. First, the global trope is not hatched out of thin air, but rather depends upon established understandings. To Fairclaugh, such rhetoric must work through existing commonsense – prevailing understandings and meanings – to be successful. The known, like an expandable turf, is seized and stretched in ways that build new understandings but as "stocks of knowledge" inseparably tied to prevailing "circuits of knowing" (Mikhael Bakhtin's (1981) intertextuality). Rhetoric thus moves back and forth between the realms of what is understood and not understood in a kind of hyper see-saw, extending the sense of what we need to know by embedding it in what is already known. New assertions may seemingly break with tradition and common conception, but always mesh with established meanings and comprehensions.

In the second key point, mutually supporting thematic "sub-stories" built into the global trope are the core of this formation. The formation, like others, is thus composed of interconnected webs of sub-stories (each a central knowledge claim) that are anything but neutral. Each, in turn, invokes a script of characters (e.g. good guys, bad guys, threatening people, community salvationists, ominous beings) and plots (e.g. the declining city, the decay of civic morality, the new dangerous globalization) to align a sub-story with established sensibilities and values. The rhetorical formation of globalization, then, makes and blends these together, giving it content and clarity. For example, presentation of a new global business person is a sub-story fitted into this formation, with other ones, that helps clarify and normalize a rhetoric of new global times. This being gains meaning and coherence by being situated most centrally within a staid and helpless locality, the reality of collapsed barriers to trade and business relocation, and the rise of a new cultural ethic of placelessness in America.

The chapter thus details the existence of three prominent sub-stories in this global trope: the rise of a new placeless entrepreneur, the existence of resource profligate inner cities, and the ascendancy of progressive mayors as city salvationists. Each is a crucial piece in the creation of a coherent and acceptable snapshot of a global reality and how cities should best respond. They are not the only sub-stories that swirl through this rhetoric, but are the most prominent and important. These, in unison, attempt to, and often do, dominate the public's field of discursivity about city circumstance and immediate need across the rust belt. Through this, the plight of Chicago, Philadelphia, St. Louis, Detroit, and other rust belt cities are to be both understood and acted upon. Mastery of the common discursive, Fairclaugh's key ground of seizure that makes a rhetoric work, has been the goal.

A caveat to all of this needs to be mentioned one more time now. It is, I believe, too simplistic to reduce the post-1990s construction of the bedraggled, civic-afflicting rust belt black ghetto (or the other "objects" crafted that are now discussed) to just the clout of the global trope. This would be an oversight that fails to recognize other political, economic, and cultural forces at work. For example, the post-1990 constructing of this black ghetto has also been fueled by the decades-old desire to sustain the sense of a racialized "other" that has bolstered the imagined authenticity of "whiteness" and a desire to spatially manage "a people" borne of stereotype and fear, among other influences (see Hooks 1993; Wacquant 2002). Thus, the global trope is best seen as the dominant, but not exclusive, influence in this constructing. The global trope, as a discursive unity, becomes a key constructing apparatus by speaking eloquently about many things (i.e. direct thematic utterances, staging the veracity of these utterances) but operates amid the flow and flux of other influential rhetorical formations. That said, I begin my deconstruction.

THE NEW SPACELESS ENTREPRENEUR

A key element of this global rhetoric has been the tale of a central figure: an offered new footloose and place-detached business person that now dominates local economies. This discussed being, a projected subset of the broader business community, is narrated amid an elaborate set-up that features a new reality of investment and business hyper-mobility, a cultural ascendancy of placelessness in American society, and a now shrunken globe marked by constellations of global interactions. This frame, like an ensnaring cage, discursively draws people into a projected world that provides meaning onto the identified subject, this new footloose business person. People are nudged to confirm the reportage of this business person by recognizing a supposed new reality that suffuses "him." The depiction of a disturbing and unsettling economic being, in this case, is inseparable from renditions of a new, harsh economic world.

To "make their man," these narrativists widely use the form of melodrama (see Walkowitz 1993) and the dense cultural content of place-destructive, anti-community operative (see Eagleton 1991). These forms, stylistic bases of meanings and understandings, mediate the presentation of the new world (the frame) and the new business person (the central subject) as the "truth" of drama, spectacle, and emotive reality. While melodrama is a flexible mechanism to code the world, the anti-community operative construct is a malleable but razor-sharp cultural form. Both have a long and deep history in the U.S. experience of being used to construct diverse characters (e.g. immigrants, the poor, foreign investors, inner city African Americans, and bankers and industrialists) (see Schneider 1988; Laws 1997).

Through these forms, these business people are set out in riveting and pejorative terms: their actions, non-stop and fundamentally self-serving, shake neighborhoods, weaken business districts, and decimate jobs bases. At the core, fickle decisions can close factories and generate destructive waves of unemployment, relocate plants and facilities and annihilate neighborhoods, and steer investment elsewhere to punish local tax bases. Here are tactical bodies, always thinking and scheming, that reduce a lived world – the city – to a simple opportunity structure. They, plugged into the circuitry of global operations and the bottom-line realities of a cash-register mentality, funnel all that they see and touch through the lens of economic functionality. What their actions threaten, in the final analysis, is the projected heart of city civility: cultural fabrics, economic stability, and social life. Thus, the making of a civic menace proceeds through colorful and evocative text (see Norton 1993).

In theme, a problematic reality now looms over the city. These global business people, so important to the sustenance of the city, cannot be domesticated and tamed: any appearance of this is just that, an appearance. A kind of economic savagery compels them, saturating their motives and actions, that is entrenched and resilient. Beneath the cover of rhetoric, a simple, irreversible reality lurks: a calculating, economic creature that acts in its own self interest. To forget this, as Ambassador Felix C. Rohatyn (in Scrimger and Everett 1999) notes at the U.S. Conference of Mayors Meeting, is to "ignore . . . [how] globalization has obliterated frontiers [and created new business people] . . . public officials who ignore this [new business person] do so at their constituent's peril." Now, to Rohatyn, the public has to think about something crucial: how this self interest can be manipulated to serve city interests. If it was economic climates and lax regulations that these entrepreneurs really want, perhaps this is what the cities should be providing.

But there is something different in this rendition of an urban demon. For how could these narrativists so flagrantly excoriate business people who have been powerful, visible, and about-town beings? As with Jamie Peck's (1995) "movers and shakers," many of these people have anchored these city economies. Because in all its clarity and luminousness, this depiction is of an abstract archetype and everyone seems to realize this. This caricatured economistic person, signaling a barely identifiable but scary being, communicates a fear more of a pervasive process than a particular person. Renditions of global business people, as the pages ahead show, have been relentless yet de-personalized and are seldom affixed names. These beings, in this sense, do not communicate a careful rendition of a person but signify an emotively codified reality. They are thus to be known, above all else, for the processes they represent. Yet, for what they represent amid this mix of clarity and ambiguity, these business people easily become unforgettable and resonant: no figure is more central in the narrative and crucial to its content.

Thus, the construction pivots around the dual tropes of "exposing" an impulsive chaser of profit and making "him" an increasingly dominant local business person. Bold oratory, embodying both tropes, declares the problematic footlooseness of a numerically ascendant, culturally rootless being nominally tied to place: his quest for profit is relentless. The result is a ruthless economism and an unremorseful predilection to out-migrate. Thus, Citicorp chairperson Walter Wristen (1994) offers the meteoric rise of a stark new urban business person to New York City Mayor Rudy Giuliani: "the advent of the information Age makes capital even more mobile than in the past; it will go where it is wanted and stay where it is well treated." As brute economic beings, to Wristen, this new breed of business person cares little about place and "will flee onerous regulations and high taxes . . . New York has to create a climate to attract capital. The high-tax, overregulated environment you inherited repels capital."

At the same time, the script frequently deepens the ominousness of these entrepreneurs by making them posturing and flamboyant beings. Assaulting their form, they are said to wittingly wield a cultural charisma, coercively bandied around town for public consumption, which provides them a polished, benevolent appearance. Thus, when cities are deemed useful, these business people are involved and visibly appear as humble civic servants. Participation in civic groups, charities, and city planning in this context is common and effortless. But just as easily, when the lust for profit sweeps over them, this frequently disappears. In a flash, commitment to place is often unapologetically dropped. The result, as Cleveland mayoral candidate Jane Campbell (2003) notes, is a sobering reality: "here is the cold truth: we are facing a second revolution . . . technology and capital are [whimsical and] global . . . In this world, our only true unique resource is our people and their skills . . . As capital and jobs flock to the booming suburbs, the challenge of balancing economic development throughout the region will grow". Campbell's capital is hyper-mobile and can run roughshod across cities, they are fundamentally about making money.

These narrativists also frequently offer a sordid twist on this identity: they forward a perversely playful business person who often joyously partakes in a kind of game of musical places. In the presentation, these are bright, energetic, and ruthlessly mobile people drawn to the perverse pleasure of maximizing profits. The lure of brilliantly playing one community against another to find optimal business climates intoxicates them. Workers, neighborhoods, cities, and local ways become discardable items that serve as kinds of toy-pieces to these business people. As Cleveland Planner P. Reid (2004) noted, companies and corporations today commonly shop around and look for good areas to invest . . . they want tax benefits, land deals, infrastructure subsidies . . . in Cleveland, the CEOs and Heads I've talked to want best locations, but for many of them it's also a kind of game for them, you know, kind of the thrill of the hunt . . . it's the new entrepreneurial culture today . . . it's all over the place and you can't prevent it.

Serving up perversely playful business people is potent symbolization: it displays revealingly recreational beings. But this is not an isolated usage of this metaphor. Creating threatening beings, to Hooks (1993) and Mercer (1997), often involves this trope. It is used to display the "real" of people (i.e. their values, beliefs, and concerns). Not surprisingly, then, numerous late-twentieth-century urban folk-devils constructed in America – black youth, Hispanic teenagers, black men, welfare mothers, and immigrants – have been vigorously narrated through the lens of recreational pursuits (see Mercer 1997; Kelley 1997). Thus, black youth in neoliberal presentation ritually participate in "wilding" for fun, poor black men commit to lives of hedonism, welfare mothers zone out on sci-fi flicks or embrace Welfare Queen lifestyles. Such styles and characters of play are the mirrors set-up to reveal the supposed truths of a people's values and beliefs (e.g. such people are culturally off-the-map and in need of social revamping). Similarly, revealing the playful side of these rust belt entrepreneurs is an effective ploy, which displays for all to see the supposed true core of a problem being.

However, a caveat to this is that all is not so negative or simple. In a vivid contrast that reflects this rhetoric's complexities, this being is also frequently projected as talented, disciplined, and perversely admirable by these same narrativists, often in the same texts. These more complex and hybrid damn-the-new-spaceless entrepreneur texts, also, discuss a business person who has seemingly mastered the craft of efficient production, works unceasingly to pursue profit, knows the outcome of all conceivable moves, and adroitly maneuvers in the rough and tumble business world. To understand this paradox, it must be recognized that these are business people with the same neoliberal values that the growth machines extol. Projecting them as negative is tempered by the desire to uphold something basic: their value structure. Typically, in the prevailing neoliberal rhetoric in rust belt cities, sound business people are cast as "the real shapers of our times, active in government . . . cities . . . cultural industries . . . the change-agents, innovators, discoverers, and alchemists [that] reinvent products and services" (see Friedman 1995; *Carnegie Mellon Alumni Magazine* 2005).

This tension in renditions of the new problem business person is resolved in two ways. First, some narrativists serve up civically damaging but rational and understandable "global wheelers and dealers" whose actions are defensible. Their actions are presented as community damaging but entrepreneurially-logical acts. These actions, economistic and city-ravaging, nevertheless follow a kind of iron-law of capitalist economics (e.g. choosing optimal places to produce) that make sense. Business commentator David Brady (2005), for example, speaks first of a downside to the actions of these new global players that needs to be dealt with. Cities, as vulnerable places, must have initiatives to meet them head-on. To Brady, losing the likes of small and middle producers, let alone large producers, is cataclysmic. But, Brady also speaks of the reality of "economic science," with profitability pursued across spaces as "managers are agents . . . principals are the shareholders, and . . . the rule governing behavior is that the managers should maximize shareholder value."

These narrativists, not surprisingly, frequently lapse into a relatively charitable rendition of these business people. Here the profit-above-place, city-indifferent creature is supplanted by a more civic-invested, open-ended "guy." This business person's predilections, now, are said to follow a flexible realism about concerns for community and business that is anchored in a reality of unforeseen, ever-changing entrepreneurial situations. A kind of "pragmatic dance," rather than a rigid accumulation blue-print, marks their decisions and actions. The need to ceaselessly innovate, not narrow self-aggrandizement, is the projected dynamic behind their actions. As planner M. Field (2004) notes to us,

> the new breed of businessperson [in the new global times] feels for the needs of the city and the people, I think . . . but for them the bottom is still where the money can be made . . . These businessmen [sic] live in and engage the institutions and people of St. Louis . . . but they can never lose sight of the bottom line, that is, I mean, where the company can best flourish.

Other presentations resolve the tension of the spaceless entrepreneur in another way: by isolating and fingering only their personal greed. These people, cast as extreme economistic creatures, are aggressively moved into the realm of outrageous and bizarre and narrated as idiosyncratically place-detached and locally insensitive. This villainizing, for all its hubris, ultimately protects the new neoliberal belief system: it establishes a person whose bizarre indifference to place puts them and not their guiding imperatives out of the cultural and civic mainstream. Personal idiosyncrasy, not laissez-faire markets or fervent government support of capital, is the issue. As one St. Louis Alderman noted to us:

> globalization today is being fueled by too many businessmen who have gone too far . . . they care nothing about where they're conducting business. I see this in St. Louis . . . And when that's the case, they care little about the city and their communities. Being footloose and globe-trotting wreaks havoc . . . as they move plants and businesses with little remorse.

Yet this offering of entrepreneurial problem person is not new in America. Decades of expositions in certain historical periods narrated ominous entrepreneurs who followed crude profit-taking and hurt communities. In the 1920s, for example, newspapers and magazines widely depicted exploitative and amoral business people who would employ any practice (selling shoddy merchandise, promising untenable warranties) to stay viable (see Noll 1990). This display, typically crude and afflictive, asserted one easy-to-grasp understanding of the Great Depression that implicated agents and agency rather than structural economic forces. Even when structures were implicated in the mainstream media, their impacts were typically softened by the dual presentations of causative, seamy business people (Boyer 1983). Moreover, in the mid 1940s economic malaise, bizarre economistic entrepreneurs were again trotted out to provide the public with a sense of cause and agent involved. Uncertain economic times during and immediately after World War II, Studs Terkel (1970) notes, found a receptive public to this.

The post-1990 situation that sculpts these rust belt business people is different and the same. On the sameness front, the plight of rust belt cities again calls out for easily identifiable people and processes that can be fingered for city woes. These cities continued to suffer in the mid 1980s and beyond from a litany of things: tax bases continued to stagnate, downtown gentrification was spotty and set against miles of dilapidated neighborhoods, and closure or out-migration of plants and factories persisted (see W. J. Wilson 1996; Brenner, Jessop, Jones and Macleod 2003). At the same time, the explosion of the low-wage service economy pushed hundreds of thousands of workers into this sector due to the disappearance of manufacturing (see Marable 1997; Ehrenreich 2002). Growing human misery, failed policy forays, and broken politician promises of greener pastures needed scapegoats; the ascendant global rhetoric has offered an easily digestible villain.

On the difference front, the post-1990 era has its own distinctiveness. The maturing of recently ascendant forces, particularly neoliberal policy and revanchist sensibilities, provides growth coalitions with an opportunity to re-assert a "politics of resource distribution," re-make the city in their own image, and in so doing, replenish, in particular, real-estate capital. The 1980s Reagan era set in motion rhetorical formations and a new organizational network (i.e. state programs, policies, institutions) that could be deepened to capital and the state's benefit. In this setting, the 1980s neoliberal rhetoric was thick. A new cast of anti-civic characters – greedy global business people, inner city welfare mothers, black kids, generic welfare chiselers, and immigrants – were aggressively narrated. This cast of undeserving beings established a general strand of feeling in the common consciousness that the 1990s global trope in the rust belt now works through.

A final point about the construction of this new global business person is that key rhetorical props are used to bolster this offering. Two are most notable: serving up the imperiled future city and the new global economic space. First, the imperiled future city is an evocative and shock-inducing mental map. Producing this imagined space, integrated into the linguistic text, seizes "readers" and takes them down one narrow path to understand the current drama of their cities and these business people. This discursive cityscape reveals in easily digestible form what could follow from this business person's actions: a future de-stabilized city that is already fragmented and threatened in new global times. This compelling map, shooting in and out of discussions, flickers across the linguistic terrain of the narratives to transmit a simple but potent theme: an impending urban crisis. Here is Cleveland State University's (2001) future city of "population and income growth no longer driv[ing] it . . . and the economy is not growing . . ." It was Milwaukee Mayor John Norquist's (2000) soon-to-be city of "giant industrial problems . . . no buildings you could put on a post card . . . park[s] where nothing works."

This discursive cityscape is ultimately a disciplining visual regime that hangs together in an elaborate production of space. Space is made by agglomerating in one setting the notorious icons of "deteriorated neighborhoods," "downtrodden people," "tired and declining parks," and "eroding downtowns." City image is fostered by the blurred semiotic interconnections of relational objects etched into a fabric (i.e. these characteristics are mapped onto this space to infuse it with a texture of meanings that continuously reference and draw strength from each other as metonymic associative chains). Each is a relational signifier that is semiotically grounded in this chain. Downtrodden people are set in deteriorated neighborhoods, these neighborhoods sit beside eroding downtowns, these downtowns are ringed by tired and declining parks, these parks are inhabited by downtrodden people, and on and on.

Second, the new global economic space enables this projection of a scary, placeless business person by providing the new expansive arena of possibility for business. It offers something textually crucial, the all-important ground that this being is projected to navigate, negotiate, and be guided by. Yet, this offering, while resonant, is curiously vague. It features oceans, continents, and countries that have supposedly collapsed on themselves and have opened up a limitless economic playing field. But it provides little depth to the causes and current concrete impacts of this new expansive global terrain while offering a sense of ominous threats and possibilities. Businesses, reduced to frenetic and fluid dots set against expansive global flows, signal a new economic world. The public, taken to a world's eye view and implored to scan across its vast terrains, are to imagine what Gibson-Graham (2002) calls "the telos of capital on the move" and think about the new possibilities for local business.

The resource of scale turns this offering of new global economic space – like the offering of the imperiled future city – into a strategic rhetorical device. This increasingly recognized resource in rhetorical formations is an imposed mental map that organizes one sense of the world and banishes others to oblivion. Thus, scaling of social issues defines "spatial fields of comprehension" to understand the world, which are human impositions rather than natural metrics. Such constructed fields permeate all social issues as they are debated, struggled against, and narrated in different ways, for example, the geographic identifying of the current economic circumstances of cities, street crime, and inner city poverty. A served-up "field of vision and cause" imposes a kind of conceptual reach that drives one understanding of the world to the exclusion of others. Scaled designations, seemingly innocuous and neutral, anchor the production of what Tim Creswell (1996) calls visual politics that direct a sense of what the world is, how it is composed, and what constitutes it.

This use of the global scale in the rhetoric is anything but surprising. As Herod and Wright (2002) note, this scale's power is formidable: "arri[ving] at the global position is no small achievement, for the global scale is the acme of scales and the power to proclaim the globality of any event is the power to put the world on alert." So used to narrate the new global business person, no space or person in these cities seems able to escape the new economic reality. Now, all business people and economic activities could seemingly move. Frantic flows infect all; either you move or are affected by it, there is ostensibly no escape. The world and its rust belt cities have not merely evolved, they now find themselves in a new economic order. Here is the appearance of George Soros's (1998) most recent societal "sea-change" that carries in it the potent ingredients of scale, process and anxiety.

This construction of the city-threatening business person, like other soon-to-be discussed sub-themes in the global narrative, is ultimately a sophisticated crafting. It is a profile rooted in widely held understandings, which deploys a strategic set of maneuvers (imposing presences, illuminating staged features, manufacturing absences, choreographing a clarity and coherence) to set up an expedient world and make its case. In the process, it annihilates from reality forces, beings, or processes that run contrary to a desired thematic content. But all is not so pat: because features of this rendition of character have to be recovered and re-used in the rhetoric – commitment to private markets, creating a business-oriented government, and fostering of public–private partnerships – a deliberate tempering or "extrematizing" of these attributes often occurs. Protecting the sanctity of these values as abstract societal presences is keenly desired. This constructing of the global business person, ultimately reflective of its makers' desires and ambitions, is an influential politics.

Problem inner cities

A second central thematic inclusion in the global rhetoric, equally strategic, offers resource draining and minimally contributory "inner cities" in new global times. This sub-story, identifying the city terrain that least deserves public and private resources in new global times, is also chronicled amid an intricate staging. Here, the stage is a new social and economic reality that grips the city: failing schools, an increase in youth discontent and gangs, the eclipse of an ethic of social civility, the decline of and need for more tax ratables, a decline in city economic fortunes and the need for workers to be more productive. This frame, an entrapping discursive surround, infuses the presentation of inner city with logic. Once again, seemingly objective reportage of rust belt cities is affirmed by a meticulous setting-up of the presentation. Relationality ultimately transfers meaning to the object in question without which an affixing of content would be impossible. This rendition of inner city, paraphrasing Bakhtin (1981), only appears to offer a wealth of absolutes and final form.

At the core of this theme's importance, we must realize that the global trope is as much about how cities should respond to globalization as the essence of this supposed new world. Both issues equally comprise the core content of the rhetorical formation. This problem inner city zone, repetitiously narrated as something usually distinctive from "the downtown" (the foci for capital's plans to revitalize), serves up the area that is least civically productive and least deserving of public and private resources in new global times. A logic guides this communicating: the uneven development that these growth elites desire requires sculpting and re-sculpting morally uneven cityscapes of the imagination. Withdrawal of public support and resources from areas to concentrate them elsewhere, simply stated, requires a maligning of such areas. For the rhetorical formation to work, this is less political expediency than political necessity.

This key set-up piece easily flows through globalization narratives as calcified downtown blocs of declined districts and neighborhoods. This general low zone in the mapped-out city organization is projected to bear something strategic: the troublesome imprint of economic collapse and poor Black and Hispanic living. Graphic images are affixed to this space (e.g. blanketing physical decay, deteriorated physical infrastructure, hordes of aimless kids populating street corners, and ineffectual social institutions struggling to survive). The tale is structured on sharp and static contrasts: the light and vibrant outer city/gloomy inner city; external webs of supportive organizations/failing inner city organizations; and external stable culture/internal maladaptive culture. The aura of an impenetrable cultural and class divide, woven into these narratives, fabricates a space that a St. Louis Alderman in discussion summarily called "the old, minority-troubled area of our older cities" and "the poor, aging core."

Such portraits, typically visual and spectacular, are haunting and bolster raw stereotypes. Perhaps unsurprisingly, one key to this is narrations of the age-old demonic icon, black ghettos, which move in and out of these inner city renditions as a vessel for all things inner city. These narrations – evocative of a ghetto character, morals, and cultural content – reinforce the offered content and nature of these inner cities. What the inner city is, the ghettos is, and vice versa. At the core of this, these problem ghettos seamlessly meld into the renditions of inner city to create a unified, discursively inseparable terrain: the ghettoized inner city. The borders and differences between these terrains dissolve under the relentless narration of a mutually shared assemblage of dark and ominous presences. The ghetto thus luminously travels through these renditions of inner city not as its core, nor its epiphenomenal layering, but as its inseparable, all-penetrative essence.

This offered inner city, as a multiplicity of constructions, features three prominent versions. Each, in its own way, reflects capital's privileged gaze. The first stages vivid contrasts between decayed inner cities and "vibrant" metropolitan sections (suburbs, gentrified spaces, stable city neighborhoods). This comparative theme, overt and flagrant, rigorously juxtaposes a sense of the old and obsolescent with the new and futuristic. The second rendition spotlights in detail the absolute raw character of inner cities as universes unto themselves. Colorful descriptions typically feature deteriorating factory districts, dark and foreboding streets, and struggling, unstable neighborhoods. The third rendition serves up dizzying, out-of-control decline that reflects a fantastic and sensational eclipse into otherness. These dramatically transforming spaces are presented as dynamic, processural unfoldings that have now fallen into an extreme disrepair. In each case, these are puzzling problem zones and seemingly belligerent to change. An illustration of each:

> To the enormous crisis of the inner city . . . solutions [have not been] found in dealing with the more complex pathologies of the lack of hope and economic opportunity, and the decay of cultural values . . . Every day . . . the public allows legislatures to waste their collective breath with symbolic laws that merely address the symptoms of social pathology . . . another day . . . [and] the problem grows worse.
>
> (Independence Institute writer David B. Kopel 2001)

> They are the urban areas left behind . . . most metropolitan areas have seen tremendous improvement . . . suburbs boom, grass is green . . . lack of inner city investment [leads to] poverty, deteriorating neighborhoods, aging infrastructure, youth violence, economic burden.
>
> (CNN correspondent Kathleen Koch 2001)

> Inner cities in the Detroits, Clevelands, Chicagos, and Pittsburghs continue to suffer. The decline has deepened . . . run-down buildings, abandoned factories, gang controlled streets, and spiraling unemployment dominate. . . . What grows here is what you least want to see: temp. agencies, predator lenders, and poverty . . . these things, they're found all over these areas.
>
> (Chicago Planner P. Reid 2004)

Particularly effective are the comparative renditions that juxtapose these inner cities with their antithesis: "civic-helping" spaces. More than the other types of representation, they vividly illustrate both what these inner cites are and are not. Offered "spaces of civility" spell out to "readers" the spaces that the cities supposedly need to attract investment and play global hardball: leafy historic districts, gentrified enclaves, tidy high-tech work nodes, vibrant consumption zones. This way, inner cities are directly set-up as "the other" and graphically "distanced" as civic-productive terrains. Offering these civic and cultural "highs," following Stallybrass and White (1986), permits projection of "the lows." A kind of grotesque realism about inner cities is revealed that ultimately communicates a high anxiety. The result, a profoundly relational process of social constructing, allows narrativists to serve up luminous, improvisational sites for strange characters and happenings in a disorderly space.

But while crucial to the global rhetoric, these bedraggled inner cities are typically less narrated than the other key themes. These spaces, sketched out but colorfully chronicled, are commonly de-centered when discussions of global realities and the new global business person put these center stage. These imaginary spaces, in this sense, are crucial but fleeting objects in the rhetoric, rising to the forefront only to quickly slip to the margins after they infuse the text with meaning. For example, as Baltimore Mayor Martin O'Malley (2001) notes, "[In the inner city] too many children . . . grow up knowing nothing but fear and violence." To O'Malley, "It's in their neighborhood. It's in the street. It's even in their home." But "taken as a whole, the highs have overwhelmed the lows. As a city, we have started to take responsibility for charting Baltimore's course. And our city has moved forward in many ways . . ." To O'Malley, "people around the country – and many right here in Baltimore – are discussing that what we have is looking pretty good."

Yet these renditions of inner city, not created from scratch, modify and sustain a history of similar pejorative depictions by policy people, politicians, and academics. As Miller (1973) chronicles, such voices in the 1950s and 1960s widely offered a zone of chaotic land-uses and population-swelling communities that also included colorful narrations of black ghettos. As planners, the press, and others struggled to understand the pronounced decay of early post-war cities, they struggled even more to comprehend changes in their cores. Similarly, 1970s and 1980s renditions built on this and served up something additional, a spreading zone of disinvestment and decline that threatened downtowns and CBDs (see Teaford 1990; Beauregard 1993). Dramatic racial and economic change had recently swept across core areas, this easily blamable space existed to explain the plight of cities. Both constructions served up frenetic and pulsating areas that continued to incorporate place-damaging things: new poor families, factories, and disinvested and abandoned land. These renditions, as I now elaborate, deepen in the post-1990 period.

In this context, some of these current narratives go beyond the appearance of "factual" renditions of these inner cities and openly judge them. "Over the past 10 years," to Rochester Mayor William A. Johnston (2005), "policy and community leaders . . . [are] realizing that they can no longer afford ghettos of unproductive citizens. This is not the way to build an economy." To Johnson, "this trend can only hold back New York State in the global economy." Magnifying the projected dilemma of the inner city but also serving as a crucial support piece to construct this, inhabitants and institutions are largely divested of any meaningful agency to help themselves. The inner city's current and future plight seems a done deal. Invocations of proper government intervention and the healing power of private markets punctuate this global rhetoric, but as mechanisms highlighted to uplift only the city in a new global reality (i.e. make the city more entrepreneurially taut, enact cultural upgrade) and not these projected obsolescent inner cities.

But what linguistic plays are at the core of building these three prominent depictions of inner city? They are unified in widely projecting, first, a common motif: inner cities as living beings scarred by a consumptive pathology. Usage of this metaphor to portray out-of-control beings and spaces, Anne Norton (1993) notes, has a long history in western thought. Common thought, to Norton, widely ascribes atavistic values to uncontrollable and impulsive "eaters." In this vein, these renditions across rust belt cities widely signify inner cities as primitive engulfers of societal resources. These spaces, like insatiable devourers addicted to the momentary pleasure of consumption, voraciously consume society's subsidies, services, and goods to satisfy the immediate appetite of a dysfunctional being. They "eat" and could not stop "eating." As Chicago Planner C. Fertig (2004) notes, "Chicago's inner city, what I think of as the South Side, is tied to city and federal resources that it has been hooked on . . . neighborhoods and politicians have known the resources were there and ripe for the plucking . . . if this area is to improve, weaning households and populations [here] off this is a necessity." This space's residents, caught in a crazed consumption, are problem beings.

In this maneuver, pathological consumption is decisively and potently put on "the ground." On the land-use front, advancing rows of scorched blocks, boarded-up buildings, and derelict industrial zones dominate. A monolith of insatiable decay and failure stretches across a vast wasteland. On the people front, degenerate eaters are placed everywhere: underemployed and idle men, welfare mothers, clusters of gang bangers, street people. No one here can seemingly escape a frayed social fabric that engulfs an entire area. An anthropomorphic space, made dysfunctionally consumptive and at odds with normative conduct, seems unable to be any different. Within this general area of erosion is one of the iconographic centerpieces – "the black ghetto" – frequently discussed and seamlessly folding into the offered character of inner cities that tightly bonds these elements as an inseparable unity.

All three prominent renditions of problem inner city are frequently supported by one more key "textual" maneuver: offering a decisive history of a past inner city tear (on this trope, see Wilson 2005). Widely chronicled is a supposed 1960s and 1970s cultural and economic collapse in inner cities, an unadulterated inner city implosion, as a telling historic moment that is a key supposed proof of their current eroded realities. This narrated 1970s collapse offers something resonant: a decades-old sudden proliferation of dilapidated properties, unemployment, alienated youth, decaying and unsafe parks, and obsolescent industries. In prose, the fall was swift, dramatic, and theatrical. Such use of "dramatic fall," to Anne Norton (1993), serves up a powerful ploy, "extrematization," that is ideal to frame discussions of present places. Such falls spotlight the demise of stability that helps locate a space's current condition in a key "truth" about its past. Projection of current situation, in this process, becomes effective via a deft contextualizing.

This "revealing" of past inner city tear is often done in brief but bold rhetorical flourish. Evocative terms shoot out luminous images that invoke a theater of decline. Thus, to writer John McWhorter (2002), "inner city Indianapolis in the 1970s was increasingly marked by ". . . alienation and brutality . . . black rage and hopelessness multiplied . . . this swift slide downhill . . . cause[d] . . . an epidemic of poverty . . . trigger[ing] [a] collapse of civilization . . ." McWhorter's deteriorated inner city of Indy, today, bears a legacy: a 1970s rapid downward spiral. Similarly, to the New York Public Library Digital Project (2004), "deteriorating physical and social conditions of the inner cities of the North . . . were as tangible as the economic upturn and growing job opportunities in the south in the 1970s." "During the 1970s," to this chronicler of the New York City situation, "the problem of the ghetto – urban decay, inner-city poverty, and unrest – appeared urgent." This rendition of Rust Belt America in 2004 is of an obliterated terrain with profound ties to something monumental, a 1970s downtown collapse. One version of history, wielded as a discursive prop, proves a key resource in the creation and legitimation of this current notion of inner city.

These inner cities, in the final analysis, are to be both accepted and objects of serious concern. The public is to now know what these spaces are; further interpretation is senseless. But, with integration into the body of the global trope, people are to be worried about these resource-draining and resistant-to-change terrains: they stand poised to undermine the city restructuring project. "In a new world," to Chicago Planner C. Fertig (2004), "seismic shifts are occurring and everyone and everything has to be re-thought." The reality of a brute and easily encodable globalization has arrived; it is everywhere and profoundly penetrative of local life. "School is out," to Fertig, "now it's time for us [Chicago] to act and be responsible." This inner city object is, ultimately, as important a fixture fitted into the global trope as any other theme.

New heroic mayors as salvationists

The aggressively narrated rise of a new, heroic rust belt leader, writer Dean Mosiman's (2002) "new entrepreneurial mayor," is the third major sub-story in this global rhetoric. As shown in this discussion, these beings gain meaning in an elaborate construction of their situation and world (Wagner-Pacifici's (1994) "discursive surround"). This rendition of mayor is strategically located in the narration of an economically uncertain city, a politicized local milieu, the reality of hovering global conditions, and an inherited legacy of bureaucratic welfare-state programs that stubbornly persist. This world

– its content, problems, and possibilities – makes and animates this sense of mayor as an expedient discursive context, deftly stages their choreographing. Discursive surround and central thematic utterance, as in the previously discussed sub-stories, imperceptibly meld.

Discussions about these mayors, most directly, identify the civic visionaries who should lead the needed physical and social restructuring. They are cast as emboldened and astute neoliberal beings that decisively see and know all in the city: Mickael Bakhtin's (1981) "eye of reason." A human form, made both a resonant symbol of and an articulate mouthpiece for best revitalization values and strategies, positions these mayors as key narrativists and key exemplars. Through them, the public is to respond to something crucial: a rational, omniscient voice that clarifies municipal ills and best solutions. This voiced hero, reveling in a kind of analytic virtuosity, collapses an inchoate process ("new global times") into easily understood images of a new civic problem, the solution, and the appropriate leaders to spearhead change.

At one level, the production exudes an elementary operation of power. These city salvationists are communicated as ideal beings to carry out city change: They supposedly know city needs, the realities of local politics, and who has to be stifled as obstructionists to city advancement. This eye of power sees and knows all: they are the anointed makers of city renewal and re-creation. As the Indianapolis Regional Center Plan (2001), a public–private city growth consortium noted, "the new breed of big city mayors has welcomed business back into the city, stressed performance and results at city agencies, downplayed divisive racial politics, and cracked down on the symptoms of social disorder." "As a consequence," to the Consortium, "America's . . . cities [can] become vital communities once again." There could be no better agent of change, the consortium notes, for these globally-challenged cities.

But this mayoral construct is also sophisticated to reflect the more subtle operation of power. A hybrid personality is deftly sculpted to render this desired product. Often, this being is provided a privileged place of birth, appropriate kinds of socialization, and immense real-world experience. Seizing the ground of commonsense neoliberalism, these mayors are made to bear central notions of prevailing public sentiment: the benevolence and wisdom of "localness," wariness of politics and politicians, and "down-to-earth" conservative values. Populated with such meanings, this character is injected with an unshakable populist sensibility that knows neighborhoods, local tastes, and local culture. This projection of grounded, neoliberal guy, as a creative construct, takes the public to a carrier of their own sensibilities.

Media narrations of the 1990s neoliberal political darling, Stephen Goldsmith of Indianapolis, illustrate this. He is widely depicted, first and foremost, as a locally attuned, intuitive neoliberalist. He is an Indy "guy" who, infused with grass-roots values and aesthetics, toils to make these a reality. Bearing the local in his soul, he aggressively strikes out to enhance what the sensible Indy community supposedly wants: everyday livability. To local planner B. Braggs (2004), "Goldsmith is Indy born, raised and educated . . . [and] made his mark as a leader who knew what Indy. needed and wanted . . . he re-energized local spirits and the business community with projects like Circle Centre Mall and Conseco Fieldhouse that gave us pride and national standing." Under the leadership of Goldsmith, to the Annie E. Casey Foundation (in Professional Experience 2003), "Indianapolis has fundamentally changed the way it . . . provides services and interacts with citizens." Turning to a gutsy local son, "his [Goldsmith's] ambitious and innovative reforms have helped revitalize the core of the city [and] cultivated neighborhood empowerment."

Wrapped in this veneer, mayors like Goldsmith are projected as keenly business-embracing. They are haunted by an insight that reflects their persona and defines their raison d'etre: that future urban survivability requires a tough and stern disciplining of people and place. Drawing out this scripting, they are narrated to widely traverse urban spaces and give dynamic speeches propelled by the mission to entrepreneurially discipline cities. In all their stops, local and extra-local, this action is a constant. For example, Tom Cochran (2003), Executive Director of the U.S. Conference of Mayors, is shown to spread this theme across Rust Belt America via talks in places like Cleveland, Denver, Pittsburgh, and St. Louis (see Travel Business Roundtable 2005). This projected truth teller knows these cities' plight and travels across this region to spread his message of pro-business. Cochran, in his own words, notes the reality of the birth of the "entrepreneurial mayor" and that now we must "bring corporate America even closer to the mayors of our nation" (in Mosiman 2002).

In this context, the theme's scripting is dynamic and theatrical. Many of these leaders move feverishly and improvisationally across cities as crusading figures. The space under them is a stage for their own personal drama, a place to test and exhibit their moral fiber. Visits commonly feature forceful assertions, articulate oratory, and clear vision. Placed before many audiences – business people, church congregations, city council members, the press – their commitment to a progressive, re-entrepreneurializing of cities is unshakable. Decisive speech strips away pretense and false ideas to reveal a commitment to a package of crucial core principles: the prowess of private markets, the need to "culturalize" cities, the need to get the state out of "welfarist" endeavors, and the necessity of fine-tuning investment climates. At the same time, they starkly map the social and economic malaise of their cities to provide the crucial discursive frame. As Indianapolis Mayor Stephen Goldsmith (2003) notes:

> It was precisely this type of free fall in which many of today's mayors found their cities in when they came to office during the 1990s. Our concern for the direction of our cities crossed party lines, and many of us found that despite our political differences, the isolation that cities felt from shouldering the main brunt of the blame for the nation's social

> problems was a remarkably unifying force. We saw one another as fellow foot soldiers in a war that pitted us not openly against poverty and the forces of urban decay, but also against our own state and federal governments. As we approach the 21st century, I am proud to say that we have finally [begun to make] significant progress . . .

To underscore the credibility of their identity and commentary, this progressive mayor is also frequently contextualized against one more thing: a tumultuous, resistant-to-change city. This reality, often vigorously narrated, is frequently a wild and wooly place filled with rambunctious interest groups and political operatives. This configuration, in these discussions, helps perpetuate a sense of what Thomas Sowell (1984) earlier called "the poverty elite" who adroitly construct something that was supposedly once less confused and misguided: common black thought in inner cities. Indianapolis Mayor Stephen Goldsmith (2003), in commentary, zeroes in on his sense of the core dilemma, a people-crushing federal government: "the enemy . . . is clear . . . a federal government [that] sent in a host of occupying forces: counterproductive welfare programs, public housing initiatives, and a 'war on poverty,' all of which simply exacerbated the plight of our cities." "As a result," to Goldsmith, "both literally and spiritually, our cities became empty shells of the boundless opportunity they once offered."

In this context, narratives often have these mayors forcefully confronting these politics and difficult-to-change places. Walks through barren terrain (obsolete industrial districts, inhospitable streets, decaying neighborhoods) and discussions in streets spell out the changes that have to be made. Mayor Dennis Archer of Detroit (1990s), for example, is often located in a hostile inner city speaking decisively about the realities of globalization (see Parr 1998; Kilpatrick 2002). Mayor John Norquist of Milwaukee is often placed on the streets walking and imploring kids to get educated in a new global world. Mayor John Street of Philadelphia is frequently placed before poor black youth in parks and schools as a voice of motivation and reason that struggles to shore up a frayed social structure (see Philadelphia citypaper.net 2004). All wind through a key icon, a labyrinth of decay and neglect, that deftly frames their articulations and identities.

This gritty and tumultuous city is ultimately a key centerpiece in a staged reality to illuminate and put into play these mayors as political props. An elaborately choreographed space becomes central to projecting an identity so crucial in the global trope. People, buildings, and streets written into a space contrive an expedient reality that, as a relational tool, anchors the producing of strategic meanings about these mayors. This relationalizing device never stops working to provide an interpretive code to understand the central subject matter, these mayors. Such relationality, following Christopher Norris (1982), always operates in processes of social construction, helping to define and embed ascriptions and assertions. Without this relationality, there could be no construction of this heroic mayor.

A final point about this constructed heroic mayor. To define them and underscore their ability to deliver the new restructuring, this mayor's repertoire is typically expanded to include a dexterity to mobilize a purported key figure: a progressive developer. Unlike the "global developer," in the rhetoric, this locally rooted one can transform cities committed to a place's people and health. Their drive for profit is tempered by the desire to sculpt something precious to them: places that are livable and economically competitive. Their blueprint for the city, in this sense, matches that of the progressive mayor. Like the presentations of visionary mayors, these developers assume a number of roles in the rhetoric: omniscient community mender, tireless worker, and supportive civic servant. These projected developers, key sites to illuminate the values of these mayors, reveal mayoral prerogatives in motion. A key strategy to advance one's political initiatives, Murray Adelman (1988) notes, is to serve up benevolent beings whose values and morals correspondent to one's own.

These developers, too, are commonly featured in discussions as tireless and visionary beings. They, also, are activated in a text of an easily read economic world (new global realities) filled with crude figures unable to see this reality: stubborn pundits, poverty-peddling politicians, bureaucratic state institutions, and blinded-by-rhetoric citizens and activists. Theater and non-stop action permeate descriptions: these developers endlessly comb the recesses of cities to turn around ailing blocks and districts. Prominent Florida-based developer Frank McKinney, for example, has been proclaimed by newspapers across this region as

> a true American original; one of a new breed of entrepreneur – an "on the edge" visionary who sees opportunities and creates markets where none existed . . . Armed with rugged good looks, a disarming personality, and a willingness to attempt what others don't even dream of, Frank has truly earned his nickname "the daredevil developer".
>
> (*Wall Street Journal*, in McKinney 2002)

As Chicago and St. Louis Planners P. Reid and M. Field note, McKinney is welcome in their cities anytime.

Of these benevolent developers, those locally born and raised tend to receive the most acclaim. These projected hearts and souls of the civic will, as true grass-roots people, strive to eradicate all forms of city obsolescence. Born and raised here, their commitment to place is often deep and hauntingly personal. They covet these cities as lived spaces where they have played, attended school, gone to church, and learned to be civic and moral. These strategic voices, complicit with neoliberal elite designs, vividly recollect and invoke haunting memories around a once social and economic stability that has dwindled. Key icons – robust families in ethnically rich neighborhoods, sturdy blocks of factories, orderly and colorful streets, and vibrant parks – litter these recollections. This idealized city that they invoke, a pre-global terrain, strategically forwards for all to see the ideal spatial form which now needs to be replicated.

THE RESULT

This chapter chronicles the content of this powerful global rhetoric that advances one version of city reality, circumstance, and need and attacks alternatives in a simultaneous affirmation and rebuke. Key sub-themes – the new global business person, the decrepit inner city, the new heroic mayor – give crucial meaning to direct statements of globalization's realness and the kinds of necessary civic responses. The sub-themes and direct statements mesh imperceptibly, stretching established understandings about the world and people to stage an intuitively acceptable and coherent rhetorical formation. Sub-themes are discursive fragments which, stitched together, help make this unified rhetorical whole by the threads of words, meanings, and signifiers. In the process, a human made story keenly submerges its roots in staging, interpretation, politics, and the rejection of alternatives, with it proclaimed a simple recounting of objective reality. At work is a dialectic of thought and sight, what Anne Norton (1993) calls the visual economy, that projects one sense of the world to impose political position.

This rhetoric communicates a global reality in three dramatic and compelling stages: social transgression, extreme crisis, and need for forceful counter-action. Social transgression is the new city-disruptive process of accelerated business and investment out-migration steeped in a dramatic growth in mobility and placelessness. Now, rust belt city economies and social life are being afflicted as businesses and investors routinely annihilate the once adhered to social contract they had with communities. Before, these entities tended to the needs of communities, this was the good old days. But times have changed: the reality of a new economic playing field ("the global space") enables the formation of a global business person who now unremorsefully trades cities in a game of musical playing fields. These beings now unapologetically gaze across vast continents in anticipation of potential departure and new set-up. Rust belt cities are in the throes of something stern, an economic quandary, whose roots seem inexorable and inevitable.

Crisis displays in oratory the city-punishing outcome of this social transgression: depleted jobs, shrinking tax bases, and eroding middle-class life. The reality of globalization has set in, rust belt cities now suffer from its impacts. To complicate matters, these cities are now purportedly headed down a possible path of destitution. In offering a sense of crisis, it is projections of the future as much as documentations of the current that are crucial. The imposed scare at the core of the purported crisis is palpable: job bases can further fracture, middle-class living can become a memory, and nothing would be left of value in the downtowns to hold middle-class attention. Statistics, both current and projected, are unleashed as the irrefutable empirical evidence that grounds this offering of crisis. These cities, framed as anthropomorphic entities, need a strong dose of health and vibrancy if they are to survive.

The necessity of counter-action, the final stage in the sequence, follows logically. Now this devastating globalization has to be met head on. That means disciplining city social and physical fabrics to make them economically competitive as players in new global times. In a systematic streamlining, all people have to be economically productive and civically contributory, market outcomes need centering as the instrument to shape physical form, and the state (amidst the paradoxical rhetoric of neoliberalism) needs to be aggressively involved in attracting and retaining companies and investment. This neoliberal package of demands, in short, promises to master the eerie overhang in current times – the reality of a punishing globalization – if urban collectivities can just commit to identifying and aggressively striking out against this. As we learn ahead, the consequences of successfully inscribing this rhetorical formation into local political and social life most severely damages one city community: black ghettos.

II.

Current ghetto dynamics

4 Glocal ghetto changes

INTRODUCTION

In rust belt black ghettos, black bodies are now openly stored and isolated as problematic city contaminants that are in their rightful place. Growth machines communicate that the creation of separate, class–race worlds is a necessity for a generic "public good" given something intractable: contemporary economic realities. In this vein, the two mile distance from Chicago's Loop to the South Side more profoundly separates already divided worlds. Philadelphia's North Philly, walking distance to sparkling Center City, increasingly exists in another universe. These districts are more intensely trapped in what Martin Luther King called more than 30 years ago "a triple ghetto:" a ghetto of race, a ghetto of poverty, and a ghetto of human misery. Yet, all is not so simple: even in flagrantly conservative times, as discussed shortly, this deepened marginalization must be continuously crafted and symbolically managed. It persists as a contested fault-line that growth machines, via pronouncements, newspaper commentary and reportage, and planning text, toil to control.

This chapter examines the nature of these ongoing changes in rust belt black ghettos. Three key changes are discussed. First, levels of poverty and despair now deepen in these communities. Economic and behavioral data indicate a heightening of suffering – grossly underreported or virtually silenced in the mainstream media of cities – as a throw-away capitalism more deeply discards these spaces and populations. Second, these ghettos are being widely constituted and understood in new, debilitating ways. Currently, one major description, a rhetoric of a degenerate people and space at the local level, has supplanted the text (i.e. the national rhetoric of the Reagan era) of a falling into darkness and spiraling out-of-control people and space.[4] Third, these ghettos now have an increasingly uncertain relationship with a key institution that has recently helped consolidate their existence: the prison system. This connection deepened after 1980 with expansion of the prison-industrial complex that reinforced the isolation of these spaces and their populations. But since 2001, with depressed economies and financially strapped states and cities, new prison building has ground to a halt and the logic of pursuing this course is being widely questioned across Rust Belt America.

But not to be forgotten is the crucial underpinning to this: the attempt to discipline the languishing rust belt city. In the 1980s, after a brief but robust burst of gentrification and downtown upgrading, the rust belt city fell moribund in a central economic sector: the building, managing and servicing of real-estate. As a post-war privileged economic sphere in policy, this economic circuit in these cities has always been entangled with local elite dreams for profit and prestige and general notions of enhancing quality-of-life (Boyer 1983; Beauregard 1993). In the 1980s, this malaise in rust belt cities was compounded by three forces: sharp cuts in federal funding, an unrelenting deindustrialization and erosion of physical infrastructure, and a series of brief but piercing national recessions (Wilson and Wouters 2004). By the 1990s, state and civic energies turned to re-invigorating local real-estate markets (Weber 2002). The cornerstone of this, in the final analysis, has been to play to an existing "opportunity structure:" existing gentrification and still robust downtowns.

In this context, growth machines have strived furiously to make and splinter urban space to intensify land and property values. This amalgam has strenuously built coalitions and policy, and deployed rhetoric in the service of advancing this. However, a central source of tension has been unavoidable: the age-old quandary of real-estate capital working to differentiate versus equalize space (splintering versus consolidating space) that both assists and confounds the project (I take this up later). Its fundamental manifestation is capital's drive to concentrate and expand "compatible" land-uses and populations (particularly investment-propulsive people and land) while seeking to balkanize these as districts. In this setting, black ghettos, a notorious invention of capitalism, continue their historic role as an evolving apparatus in the service of capital. While some things in these cities have changed, others have not.

What has evolved is ultimately more of the same, black ghettos continuing to pay the price for "city growth." Blacks were once horded into these spaces in crowded cities to assist a nascent real-estate capital, but most importantly to enter the Fordist industrial economy to which they contributed vital cheap and abundant labor. Today, much has changed: new discursive strategies involving diverse social, economic, and political institutions work to transform city form and especially impel real-estate accumulation and local state revenues that deepens this hording. More meticulously than before, black bodies are confined to their own universes moored in a complex of inferior schools, decrepit homes, isolated social spaces, and glaringly underfunded institutions. The battering of this space's and population's materiality and symbolic character, now routinized and rigorous, engenders deeper poverty and illuminates a counter-civic terrain. This said, I amplify my points.

DEEPENED DEPRIVATION

These ghettos in the eras of anointed Super Mayors like Bloomberg (New York), Daley (Chicago), and Peterson (Indianapolis) now experience deepened deprivation. Post-industrialism continues to erode manufacturing bases and, in the process, low and moderate income neighborhoods. Between 1990 and 2000, more than 800,000 manufacturing jobs disappeared in rust belt cities. America has lost more than 1 million jobs since Bush took office (Maharidge 2004; U.S. Census Bureau 1990 and 2000). Equally important, however, is the current global trope that, first, diverts public and private investment from these spaces for their use in more "civically important" enclaves, and, second, represents them and their populations in new, debilitating ways. Mayors now supposedly fight gallantly to mediate a new civic menace – irrepressible globalization – that situates them to exuberantly speak "truths" about current city needs (re-entrepreneurialize city form, identify and re-engineer the city's unproductive, upscale city culture). City survival, now, is said to require something fundamental: unsheathing the new competitive urban form and liberating private enterprise.

Cleveland's black ghettos illustrate this reality. These spaces now reflect a city loss of more than 100,000 manufacturing jobs between 1990 and 2000 (U.S. Census Bureau). This trend continues: 62,403 jobs were lost between 2000 and 2003 (State of Ohio Council 2003). The solid base of manufacturing facilities that once lined Lake Erie and its environs – LTV, White Motor, Birmingham Steel – has dwindled. But the city's decision to shift resources to culturally upscale and re-entrepreneurialize the city is equally important. In the last nine years, more than $350 million in public funds have helped build the city's gleaming "subsidy lane:" the Rock and Roll Hall of Fame, Browns Stadium, the Old City historic area, and swaths of gentrified neighborhoods (Hennepin 2004). At the same time, public funds now flow to underwrite land purchase and new construction costs along Lake Erie. Corporations like Ritz Carlton, Tower City, and Key Corporation Center became the models early on, receiving $3,262,160, $7,920,560, and $26,370,814, respectively in abated taxes (Nader 1997).

Near fanaticism has come to characterize this drive to globalize Cleveland. The Civic Task Force on International Cleveland, formed by Mayor Campbell in 2001, quickly set out to sloganize the new initiative. Much debate centered around the best choice of slogan: Wake Up the Sleeping Giant: International Cleveland; One World, Cleveland; or Cleveland: the World's Hometown. Spirited discussion, in a serious vein, had previously dismissed as a finalist for best slogan: Where Global Opportunities Don't Knock – They Rock-N-Roll (see Civic Task Force on International Cleveland 2003). The goal, now, is to cultivate a taut, entrepreneurial-polished downtown that, to planner B. Hennepin (2004), "will be the seedbed to bring Cleveland back . . . in the new competitive reality . . . re-store solid middle-class neighborhoods, renew community energy and vibrancy."

Mayors Michael White and Jane Campbell have pulled $65 to $70 million from its low-income neighborhoods (reducing housing expenditures, job creation and training initiatives, and social service provision) to nurture this development vision (Hennepin 2004). Global rhetoric has propelled this ghetto isolationist imperative, relentlessly casting these areas (particularly the East Side) as culturally devastated and economically non-contributory. These areas, communicated to embody an all-but unabridgeable cultural and economic ethos that hinders civic upgrade, now supposedly suffer from the likes of too many "ill-willed" . . . "problem people" . . . "concentrations of broken-down families" . . . "brooding kids" (Cleveland Planner S. Plann 2004). These incendiary spaces, commentator M. B. Matthews's (2005) "third world city," are deemed to violate something important – the socio-moral foundation of a city struggling to mobilize around an entrepreneurial ethos in a new ominous global reality.

This rhetoric works through and ultimately consolidates Wacquant's (2002a) two imagined communities embedded in post-1990 common thought: the virtuous and civic working and the undeserving underclass of culturally tainted beings. Now, an entrepreneurially grounded public disgust against "a people and kind of neighborhood" (S. Plann) is tangible. Thus, two Cleveland City Councilors recently charged in bold oratory that an economically unproductive and irresponsible [Black] Eastside receives too much public aid; and the city's main newspaper (the *Plain Dealer*) declares that improving this area should center on one strategy: imposing the renewing influence of "black middle-class values" (see *Cleveland Plain Dealer*, 2005). A discussion with local planner B. Epps (2004) reflects this sentiment:

> yes the East Side is a problem, it has . . . too many welfare families, problem kids, and struggling people all in one place . . . they strain the fiber of these communities . . . today, this area needs to be proactive and have responsible civic citizens . . . and help themselves . . . Cleveland's being damaged.

Now poverty intensifies in its three most impoverished black ghettos, Hough, Glenville, and Collinwood. More than 70 percent of these households unofficially live below the poverty level (Rentgen 2004). These neighborhoods today have infant mortality rates above 15 per 1,000, a figure that rivals Uruguay's 17 per 1,000 and Mexico's 20 per 1,000 (see CIA World Fact Book 2003). On 105th Street in Glenville, only 15 minutes from Cleveland's vibrant downtown, almost every storefront is boarded up. In Collinwood, beggars and the homeless multiply across its main thoroughfare, 152nd Street, in a desperate fight to survive. Hough, one of the six poorest urban neighborhoods in America, had more than one in three of its residential parcels tax delinquent in 2001, compared to the city's less than one in ten (Center On Urban Poverty and Social Change 2001).

Equally devastating, social service agencies, which once relied on public funds, severely contract or disappear. Indicative of this, the Collinwood Community Services Center, a major community resource decimated by funding cuts, now struggles to provide basic necessities to residents: meals, day care, and housing assistance (see Naymik 2003). A crushing debt, $300,000, threatens facility closure on top of already severe cutbacks. A critical question in the eyes of the community is the agency's survivability. "At this point, I don't know" [about surviving]," says South Collinwood Councilperson, Roosevelt Coats, in a recent interview. Cash is needed to cover basic operating costs: outstanding utility and food bills, salaries, and maintenance. Desperate measures to stay afloat are now routine. "I offer creditors fifty bucks, whatever they will take, even a dollar," says Executive Director Wallace Floyd. "I'm honest and tell them that [with cutbacks] we just don't have the money."

Cynicism now marks the outlook of another social service agency, Homeless Services, the leader in combating homelessness on Cleveland's East Side. S. Phelps, a worker there, states that "very few of [our] goals have been realized . . . the situation is now . . . really . . . between desperate and out-of-control." To Phelps, poverty has deepened across much of the East Side and once solvent resource providers have shuttered or barely survive. Similar comments, from the Head of Cleveland's Domestic Violence Center, note: "It's not that we've stopped trying, but conditions are much more difficult now." To this head, "our funding has diminished, the demand for our services has skyrocketed, and we're down to a skeletal staff." "Few options remain but to go on," he adds, "but it's like swimming up a raging river [sic] . . . If we could find funds, then the real core need – creating jobs for working people – could finally get the attention it deserves, and things might be different."

At the same time, deprivation in these neighborhoods is deepened by the ascendancy of Katz and Jackson's (2004) localized, parasitic economy. In clear daylight, above-board businesses that thrive on spatially immobile, low-income customers dot the retail landscape. A cultural knowledge of situational hyper-profitability, borne of expedient economic realities and an intensified glorifying of entrepreneurial creativity, ties business actions to the specifics of a space that characterizes even multinational chain-store operations here (e.g. McDonalds, Taco Bell). This economy's heart – payday lenders, check cashing outlets, tax refund advance firms, used car dealerships – routinely charge exorbitant interest rates on loans and goods (Skabec 2004). These economic practices also characterize the more commonplace establishments – grocery stores, supermarkets, clothing outlets, gas stations – where price markups exceeding 200 percent have become ritualized and accepted (see Katz and Jackson 2004). In this setting, most common items purchased – milk, coffee, soda, t-shirts – bear the most pronounced mark of this retail squeeze.

In sedimented form is Cleveland's version of a national trend: the exploitation of a city's most vulnerable population pummeled by stepped-up marginalization. Thus, credit card interest rates nationally range from 9 to 17 percent, but payday loans carry an average annual percentage rate of 474 percent (see Katz and Jackson 2004). For Hough, Glenville, and Collinwood residents stuck in this ghetto environment, the move out of poverty or financial hardship is exacerbated by this reality. Institutional supports become institutional plunderers, the leading edge of a now coalesced predatory local economy.

As one Hough resident noted in discussion, "yeah pretty much everyone knows how these places operate . . . at least I do. But they're the only game in town . . . and sometimes you just need cash or something else quickly, or you need wheels." Another resident noted to us: I'm not really that aware [that] the [interest] rates are so high . . . it's the way it is around here, how the economy works. Why do I go to these places? It's here man, it's here, and it often keeps me goin."

The story of a devastating global trope is much the same in Philadelphia's black ghettos. The city lost more than 6 percent of its jobs (–51,000 jobs) between 1990 and 2000 (Delaware Valley Regional Planning Commission 2002). Between 2000 and 2003, this number was more than 26,000. But equally crucial has been the shift in resources to transform Center City. Wonder Mayors Ed Rendell and John Street have proclaimed a problematic malaise of the downtown and city, and have aggressively struck out to make them internationally attractive and competitive. In short order, Center City has experienced a demographic cleansing of public spaces (i.e. forced removal of the homeless and minority-young), frenetic designation of new historic districts, and a wave of publicly assisted new development (e.g. refurbishing retail anchor John Wanamaker, revitalizing the East Chestnut Street Corridor). In the process, new condos, coops, and apartment buildings have sprouted up across Center City, often in skyline high-rises or fortressed cul-de-sacs and around rigorously policed open spaces (see Reinheimer 2005).

Perhaps most bizarre, the drive to entrepreneurialize Center City now extends to something once barely imaginable: a massive wiring of its public spaces (at great public cost) to allow internet access. The city completed the country's first wireless public park in 2004, Love Park, to make its public spaces business-usable (at an initial cost of $10 million with an additional $1.5 million yearly for maintenance costs, see Kho 2004). This "first . . . and world famous economic park" (Mayor Street, in Innovation Philadelphia, 2004) provides free high-speed public internet access to residents, visitors, and businesses. Every Friday from noon to 1 p.m., on warm days, "digital fellows" walk the park to assist users. [The] Park is now more than a place where people can come to eat lunch," Mayor Street proclaimed (in Innovation Philadelphia, 2004), "it's a place to do business." It is, in his Chief Information Officers words, "a great economic development tool." Wiring of public spaces, now, is being extended to other city parks: Washington Park, Franklin Square, Rittenhouse Square, and Logan Square.

Meanwhile, funds to assist poor neighborhoods have vanished (Plann 2004). A range of initiatives spanning health and welfare, housing, and social services have been dramatically reduced: the City Year Philadelphia Program (an educational initiative targeting at-risk youth), the Treatment Institute Program (which disseminates drug treatment strategies to schools and neighborhoods), the Emergency Repairs, Preservation, and Weatherization Initiative, the Community Economic Development Program, and Employment and Training activities (see Weinberg 2003; McLellan 2003). While parks and public buildings are meticulously wired to "tele-upgrade" the city, funds for schools, housing, and food are cut. Computers, telecommunications, and upscale culture are serviced across the city, unlike depleted job bases, hunger, access to health care, and burgeoning homelessness (on these, see Independent Media of Philadelphia 2004). Philadelphia, in its leaders' rhetoric, struggles to compete for global resources that will determine its survival: that is where its resources and energies must go.

Now, two of its most impoverished black ghettos, Fairhill and Hartranft (the North Side), experience more extreme deprivation. Under new funding priorities, block grants in these areas have declined by more than 30 percent in the last three years (Bean 2004). Not surprisingly, crime rates, families below poverty level, and child maltreatment rates dramatically exceed the city average (City of Philadelphia 2004). In Fairhill, percent of poverty officially is 57.1 percent and unofficially above 80 percent, compared to Philadelphia's 6.3 percent (U.S. Census Bureau 2000). In Hartranft, these numbers are 33.9 percent (official) and 55 percent (unofficial). A casual windshield survey I conducted of both places in 2003 revealed the physical manifestations of this deprivation. Just off the main thoroughfare Broad Street, acres of boarded-up homes, vacant lots, and clusters of unemployed men line the streets. Abandoned factories also dot these areas that shelter a burgeoning homeless population with no place to go. To survive, these people regularly panhandle across the neighborhood (see Gorenstein, Boyer, and Ciotta 2006).

The parasitic economy, not surprisingly, has also arisen to dot these neighborhoods. Reputable businesses cash in on poor, immobile people in the form of "extreme stores" (e.g. payday lenders, check cashing outlets) and "commonplace shops" (e.g. care dealers, supermarkets). Thus, most of the city's 147 check-cashing establishments are in low-income neighborhoods; many are in black impoverished North Philadelphia (Brookings Institution 2005). State regulation sanctions a common practice in these neighborhoods, allowing two-week loan providers to charge an annual percentage rate as high as 450 percent (Brookings Institution 2005). Moreover, among commonplace shops, the city's rent-to-own appliance stores (used almost exclusively by low and moderate-income households) typically have installment plans whose mark-up exceeds the purchase price by 90 percent (Brookings Institution 2005). At the same time, low-income families in Philadelphia frequently pay more than $5,000 more for the same vehicle purchased by higher-income households.

These residents ultimately engage a surprisingly sturdy and adaptive economic formation. Like others across inner city Rust Belt America, it deceptively appears as a tired, staid ensemble in its last gasp of profit extraction. The reality, however, is a remarkably resilient, contingent, and taut matrix of establishments that adjusts to localized circumstances, acquires new properties, and purges others. Thus, prices for cars, food, and clothes fluctuate daily. Similarly, a widespread monitoring of customer credit-rating via formal (computer checks) and informal (merchant memory) means is continuous (Bein 2004). An exceptionally fluid inner city economy gauges its profit landscape, instills new policies of production, distribution and pricing, and generates poignant material effects. North Philadelphians, so enmeshed with this reality, come to negotiate a kind of island of differentiated economics that is always bending to bring people into it albeit in a sanitized scenario of dispassionate supply-demand realities. A population, fixed in space and time, has little choice but to "freely" participate in its workings.

In Indianapolis, the trope of globalization occupies center stage in the face of especially curious circumstances: stable job and industry growth. The city expanded its number of jobs 3.7 percent and industries 4.0 percent between 1980 and 1997 (Center For Economic Development 2004). The economy of Indianapolis, moreover, is dominated by light industry and service provision and is hardly global and footloose to which the notion "global economy" refers. Yet aggressive rhetoric about the need to entrepreneurialize city form under Mayors Goldsmith and Peterson steers government resources from neighborhood development and job creation to "culturalize" the city (e.g. build a downtown mall (Circle Centre Mall), a professional sports stadium (Conseco Fieldhouse), and acres of gentrification). Conseco Fieldhouse was built in 1999 with $79 million in public funds that dwarfs the $57 million the chief tenant, the Indiana Pacers basketball team, contributed. Glitzy Circle Center Mall, a public–private partnership built in 1995, is rooted in a public outlay of $187 million (which also provided 12,000 parking spaces within one block) (Indianapolis Convention 2003).

But Indy's poorest area, the predominantly Black Eastside, has however been devastated. New funding priorities have meant block grant cutbacks for housing and social projects of more than 35 percent in the last three years (Braggs 2004). Now, more than 60 percent of Eastside residents are unofficially below the poverty level, with under-employment rampant (see Kelly 2004). The fall-out is commonly noted by community leaders and residents. To Reverend Byron Alston, a long-term supporter of local black youth, increased crime has followed from worsened living conditions, and "we are tired of going to funerals" (in Kelly 2003). To Alston, decent paying jobs have all but disappeared, government is remote and hostile, and too many turn to illicit activities like selling drugs to survive. "We want to give hope," said the Reverend Donnie Golder of Temple of Praise Assembly, "but conditions are harder than ever." In this context, to writer Fred Kelly (2003), vague interventionist strategies by volunteers (church officials, block groups) are the principal – and a limited – corrective for the community's pressing needs.

Even momentary escape from the grinding poverty is difficult. Random walks outside this Eastside area can prove deadly. The Indianapolis Police Department, notorious for its harsh enforcement of social spaces (see Jet 1995), rigorously police the downtown area for "Eastside interlopers." To writer Fred Goldstein (2001), the police routinely . . . "harass African American youth and treat them as if they were violent gang members . . . the cops have 'jump out boys' who jump out of squad cars and swoop down on black youth."

And from our observations, police officers, black or white, move through the Eastside, paraphrasing James Baldwin, "like occupying soldier[s] in a bitterly hostile country." That is why, following Baldwin, the police ride in cars or walk in twos or threes. Poverty and blackness here, to City Councilperson Glen Howard (1991), translates into a perception of intransigent culture and values. Such "troublesome" people, supposedly antithetical to cultivating a vibrant consumer and producer center, must be regulated and controlled. Not surprisingly, Indianapolis has experienced two large riots since 1995 over police brutality (see Goldstein 2001).

Such neighborhoods across rust belt cities have subsequently become tougher places to live. In Chicago's Wentworth, a woman we talked to describes her everyday as worse than before and an ongoing struggle to hold down work, be safe, and make ends meet.

> Life now, [she says,] is tougher than it's ever been. The area's hurting, and there's no real good jobs around anymore . . . I work two jobs, both not great, over on State. I barely make the second one in time, let alone have time to change my clothes . . . But I've gotta do it, keep on keeping on, the kids need to be fed.

Another woman we talked to, living in Englewood, also notes the recent increase in hardship and declining community times. She said: "this neighborhood has grown worse . . . more hurting people, more problem kids . . . [also] my job history is kinda spotty . . . I go from one job to the next, but that's because they pay so badly." Her best job in the last two years, at a local drug store, "paid the best [but] was still too low to buy groceries regularly . . . and did not provide health benefits."

For many in these neighborhoods, anger, self doubt, and introspection intermingle. One Baltimore resident (in Koch 2001), in a common thread, blames society as much as himself for declining personal and community living conditions. He notes: "suppose we are all 'good [folk]' and have never done anything wrong. What can we expect? A large number of us out of jobs for a long period of time. Last hired, first fired, if we ever did get a job." The nature of work? "Pay the lowest . . . Type of work the meanest and hardest . . . Hours the longest at the worst time of day, or at night . . . Jobs insecure, unhealthy, unsafe." To him, the dilemma was a "white boss system . . . [that] controls . . . and controls." An Indy resident expressed similar sentiments. "It's not just us, but them," he said. "The city and society lock us up here, where it's a tough and brutal life." "I'm not sure I could live elsewhere," he said, "but at least I could be given the chance."

Deprivation in these ghettos also takes another form: high rates of health problems. The specter of early death haunts these places, particularly among the young to middle-aged. For infants in general, black infant mortality remains more than twice that of white infants (Blue Ribbon Panel 2002). In these extreme-poverty ghettos, health care official M. Spivak

(2004) estimates, black infant mortality is four times the rate of society's. Infant mortality is strongly associated with substandard and overcrowded housing, poor nutrition, and reduced access to health services (Spivak 2004). For black teens, mortality is nearly twice that of white teens in general (Tonry and Morris 1984). Most importantly, poverty and its production of deprivation, violence, and despair spawn homicide and death. In statistical terms, teens in these black ghettos face a much higher probability of death.

At the same time, both infants and teens in these ghettos suffer from much higher rates of hypertension, asthma, neural problems, and lead poisoning (Franks, Gold, and Clancy 2000; Butterfield and Larrson 2003). Two processes are crucial: the sustained producing and storage of toxic and airborne particulate materials in these ghettos and the everyday living conditions of poverty. First, these ghettos in rust belt cities continue to house or adjoin a disproportionate number of toxic waste sites, coal burning facilities, noxious chemical plants, and landfills (see Hurley 1995). As zones of political least-resistance, cities in the neoliberal era find it even easier to site such facilities here. Not surprisingly, their air, water, and land tend to be more foul, polluted, and dangerous to health. In this context, African Americans in these areas are three to four times more likely than whites to be hospitalized for asthma, hypertension, lead poisoning, and breathing ailments (see Hurley 1995; National Center For Health Statistics 2006).

Second, the sheer reality of everyday living in these ghettos exacerbates health problems (see Blue Ribbon Panel 2002). On the physical side, substandard and abandoned buildings are health hazards to kids, buildings have high rates of lead based paint, the chewing of which, even in small doses, damages brains, and air filled with soot residue and cockroach allergin exacerbates asthma. On the human-emotional side, poverty breeds tremendous stress, low incomes are obstacles to obtaining decent health care, hopelessness spawns bad diets, and deprivation erodes health habits. Everyday living here – its physical and social dimensions – thus poses profound hazards to babies, kids, teens, and adults. The everyday's unbroken flow in these neighborhoods, as the Blue Ribbon Panel reports, is rife with destructive consequences for health. Simply breathing, frequently fouled by insect particulates and excessive nitrogen dioxide from poorly vented stoves and heating appliances, can be lethal.

Thus, in New York's South Bronx, asthma and neural problems are more than three times the city average (Institute for Civic Infrastructural Systems 2004). Epidemic asthma in the South Bronx has led to its residents dubbing it the common condition "fatigue" (Ostrum, 1995). The root of the problem is the presence of a nearby waste burner (the largest in the Borough) and a cluster of power plants. The waste burner spews out noxious particulates all hours of the day – never stopping to rest – to create a "mini-snow" to locals. In the 1990s, four General Electric power plants were sited here that now blast asthma-causing chemicals across the area. The New York Power Authority deliberately sited facilities in stages here. These locally unwanted land uses were shunned via zoning in many parts of New York and located where political opposition is weak (see Gasping For Justice 2003). This "danger alley," today, hideously permeates local ghetto lives as an unseen but deadly force.

Rates of hypertension and lead poisoning in the South Bronx are also more than two times the city average (Alderman, Johnson et al. 2002). Facilitated by the South Bronx's heavy concentration of poverty, high stress, and dilapidated buildings, the effects are etched on the faces of residents in local stores, parks, and streets. To city activist G. Willis (2004), struggles to secure the basic necessities of food and shelter in the era of Workfare, No Child Left Behind, and evaporated job prospects induce the key factor for hypertension: stress. At the same time, lead poisoning is a silent terror in this area. The federal government realized its dangers and banned most uses of lead-based paint in 1977; most lead in gasoline was phased out in the 1980s. But government never dealt with the three million tons of old lead that line walls and fixtures in millions of city housing units across America. More than 75 percent of the South Bronx's housing stock, more than 50 years old, embeds massive amounts of old lead (U.S. Census Bureau 2000).

Chicago's Near South Side is much the same. With the second highest rate of childhood asthma in the country, difficulty in breathing is commonplace among kids in schools, care centers, and parks (Farella 2003). In Near South Side, the lethal mix of noxious air polluted by nearby plants and dilapidated buildings (infested by rats) infiltrates a space with an especially vulnerable population, one that has many non-insured or under-insured households. Currently, blocks of run-down buildings dot the community and two large waste burning facilities cast a pall over it. The facilities shoot tons of particulates into the air daily, creating periodic bouts of "cloud cover" that cover entire blocks. At the same time, in 2000 more than 40 percent of Near South Side's housing stock was and still is substandard, 90 percent of it being over 50 years old. Treatment of rat bites at local hospitals is among the highest in the United States (see Hurlich 2001).

High levels of stress from unstable employment and the fight for decent living conditions exacerbate the situation. Near South Side, like the South Bronx, has levels of stress that, in the words of Chicago activist D. Mung (2004). "are barely imaginable by suburban Chicagoans." Increased unemployment, under-employment, and forced participation in dead-end jobs induce stress, which is a critical component to producing asthma. In the nearby hospitals, particularly University of Illinois and Rush Presbyterian, asthma cases are a major cause for emergency room treatment (Farella 2003).

Berg (2003) and Butterfield and Larrson (2003) sum up the health crisis in these neighborhoods. Afflictions like asthma, lead poisoning, and hypertension are well understood, but continue to haunt these areas. Thus, it is known that lead paint poisoning is a much deeper problem than children eating paint chips from tenement buildings. Scientists recognize its insidious toxic effects at lower concentrations (e.g. passively inhaled paint dust, not just chewed flakes, can inhibit brain development and the reproductive system). But these ghettos continue to contain massive amounts of this toxin in buildings and homes. This health dilemma, structured by differential access to health care, prevalence of poverty and stress, and proximity to deadly toxins, is not abating. For these poor African Americans, the implications are frequently lethal. In this reality, illness and mortality become poignantly racialized and spatialized.

THE NEW STIGMA AND MARGINALIZATION[5]

Another ongoing change afflicts these black ghettos: how they are represented by the apparatus of growth machines. Frequently, renditions of impoverished African American neighborhoods, via bold but sporadic flurries, work through the fear economy to deepen the demonizing of these spaces and populations using a new regime of representation. One major strand in this, a particularly virulent symbolic ensemble, is the focus of this section's discussion to convey the strident character of this general rhetoric. This strand serves up in new ways tainted ghetto spaces and bodies, which have parallels with many previous rhetorical formations in these cities (e.g. the representational regimes of "the blighted (urban renewal needing) city," "the ghetto expanding (public housing needing) city," and the 1980s "ghetto-aberrant city") (see Tabb 1974; Beauregard 1993). Post 1990 Black ghettos once again become pawns in a broader political and economic maneuvering by capital. While this strand typically does not directly discuss the issue of new global times, it reflects the constellation of new and evolving narratives in rust belt cities (about urban issues and people) that poignantly bear the influence of this projected new reality.

This virulent strand of representation, as the most flagrant example of the new regime of ghetto presentation, narrates these spaces and populations through a mix of indifference, fear, puzzlement, consternation, and repugnance. Ghettos and their inhabitants are reduced to a kind of impossible-to-solve clinical challenge, being cast as sordid entities, objects of spectacle, and resistant-to-change elements. A diagnostic-clinical gaze looks unblinkingly at "worlds" of puzzling values and lifestyles that are tragically horrendous and sadly dysfunctional. In the gaze, ghastly cultural and social milieus decimate humanity and leave communities reeling (see Wilson 2005). In this context, these post-1990 narratives offer a sense of rationality for the public to forget the plight of these people and spaces but never to forget the sense of them as a persistent city problem. The usually unstated but widely understood relational entity to understand these spaces and people – new ominous global times – backbones the offers.

This post-1990 strand of representation does not work from scratch, but builds directly on an immediate predecessor – Reagan's "black-ghetto" oratory. Reagan's oratory, itself a syncretic production, was embraced and acted on in these cities by growth machines. The Reagan years, inside and outside these cities, featured stunning pronouncements about a supposed unpopular but necessary-to-speak reality: the ascendance of spiraling, out-of-control black welfare populations and communities (i.e. versions of the "Welfare Queen" oratory). As this section documents, the post-1990 pattern of representation deepens and stretches this via newspaper writers and editorialists, mayoral and councilperson pronouncements, and policy discussions. The focus, in full detail, is the most virulent strand of this new regime of ghetto representation that now starkly presents lost and pathological spaces and people. This strand currently shares the spotlight with less graphic accounts of black ghetto realities in these cities (which I do not take up); focusing on this particular strand illuminates this new rhetorical formation's harshest side.

The Evidence

This new virulent symbolic ensemble, like the new general regime of black ghetto presentation, appears less frequently in local newspapers, technical reports, and politician oratory than previous renditions of black ghettos. This ghetto, simply put, assumes a reduced place in common reporting and discussion of city issues. But this de-centering helps the rhetorical project in multiple ways: it helps cultivate both an indifference to and an explanation for the black ghetto situation. The de-centering communicates, most immediately, a city with more pressing needs. With this de-centering widely understood through neoliberal sensibilities, the public is to know that "minority coalitions" – their pleas and demands – are dubious and not worth extensive reporting. But also, this de-centering, a kind of cultural performance, speaks volumes about the "real reality" of this population's plight and circumstance. Bolstered by the sporadic, radiant narrations of ghetto blight and dilemma (about to be discussed), a theme is projected: these spaces and populations suffer at the hands of now excavated forces, bad personal choices and afflictive culture.

This de-centering of the ghetto object is ultimately a multi-textured communication. Now, a mobile and quick-acting capitalism barely has time to bother with these people and spaces. Social and spatial restructuring must proceed apace with obstacles to be pushed aside. Without forging a new, taut economic landscape and a supportive culture of governance, cities will decline. But at a deeper level, as the brute narrating of these ghettos recount, these spaces have to be considered and acted on accordingly. For their content supposedly necessitates the identifying, targeting, and managing of them for the public good. In discursive staging, quick time meets entrenched, problematic timelessness; an imperative to re-order must transform or safely encase an intransigent, disordered socio-spatial form that looms to threaten production of the new entrepreneurial city.

In this context, the virulent symbolic ensemble routinely represents these black ghettos as culturally destroyed. A grizzly world has purportedly driven out a raw, genuine rough-neck aesthetic of black Urban America: stagecraft has supplanted a social normalcy. Chicago's Woodlawn, to the *Chicago Sun-Times* (1996), "has bottomed out . . . [is a] downtrodden community . . . families and businesses [have] abandoned the crime-laden and drug-infested South Side neighborhood." Such ghettos, to City Planner D. Roe (2004), differ from the approximate 1980s period (the beginning of the modern neoliberal era) when these spaces were ". . . hard-working and rough, gritty places beginning to implode and . . . were evolving as very different from the mainstream." In an important way, this is a new specification of the black ghetto as an empirical object. These spaces, in the casting, have supposedly followed a twisted path to become a new and distinctive thing: a fully deformed milieu. Once falling into an abyss, this has purportedly been fully realized. The melodrama of the fall, now, is replaced by the "reality" of a fully disordered everyday that has normalized urban ills: gang-torn streets, dysfunctional families, graffiti-splattered public spaces, hustling survivalist mothers, and the like.

In the process, these ghettos are widely narrated through a metaphorically infused theme: animate places plagued by consumptive degeneracy. These spaces are communicated to be living beings fallen into a state of habitually "eating" societal resources as uncontrollable engulfers of goods, services, and subsidies. Representational styles for portraying people and places, Kobeena Mercer (1997) notes, change over time and deploy different supportive metaphors. But here, more than a different stylistic metaphor is at work: this new usage helps communicate the new theme of now hopelessly consumptive people and places. As Chicago Councilperson N. McGrath (2003) notes in discussion about Chicago's South Side: "[the area] drains the city's resources and good will . . . It . . . asks for more and more [resources] as it struggles to survive . . . subsidies and expertise are provided . . . but seem to have little effect." These black ghettos, like hedonistic and out-of-control organisms, mindlessly consume with little remorse.

Both spaces and people take a beating in the representing. On the people front, melodrama grafted to the fear economy widely features caricatured people who garishly struggle to survive in an imploded social structure. Typically, youth and adults are presented as cognitively or morally injured and range in character from pathetic and forlorn to defiant spokespersons for a new cultural order. Sometimes, as enabled "talking heads," these people in the narratives are voiced to personalize their state of disarray. But more often, they are rendered passive objects to be discussed and understood. These projected ghettoites, once widely depicted as struggling and rapidly falling into chaos, are now routinely ravenous and decimated. Before, these people hustled leading with a bravado and manhood, now they are too often angry, embrace a discordant counter-societal lifestyle, and atavistically consume. The black ghetto, in short, had devolved into a cultural wasteland that bounds and blinds residents: an angry and unremorseful devouring is the new impulse that projects an entire people forward in daily life.

The narration, also staged on the areal front, dots these terrains with the physical end-results of an out-of-control consumption: "abandoned houses," "run-down parks," "homeless people," "deteriorating buildings," and "derelict properties" (see *Crain's Cleveland Business*, 2005; *Cleveland Plain Dealer*, 2005). This assemblage of signifiers, placed in the rhetorical formation, forwards the sense of a once vital-to-community set of resources which, through grounded social and economic doings (i.e. uncontrolled consumption by out-of-control people mired in dysfunctional lifestyles), have become scarred and destroyed. "A people" here destroy a neighborhood's foundation by "eating" a litany of things endlessly and mindlessly: "government subsidies," "civic energies," "crack and other drugs," "vile T.V. images," "welfare worker energies," and "blaring rap music" (see *New York Times*, 2005; *Philadelphia Inquirer*, 2005). In communication, an uncontrolled and destructive devouring of "a people" leaves in its wake ravaged neighborhoods.

This presentation of consumption is ultimately vilification. These ghettos are starkly condemned for a complete descent into consumptive disarray that projects a deep and entrenched disorder. Black ghettos, in this motif, have been hopelessly seduced by the superficiality of immediate gratification to create a new world that pivots around shortsighted desires. Here, in asphalt and grass, is a metaphorical "false consciousness" that ensnares neighborhoods. Like lost and dazed creatures, these people and spaces continue to mindlessly consume. They know little else and, befitting low-class, racial organisms, live for the moment. This representation, a powerful politics, ultimately converts an inanimate materiality into something sinister: a living, brooding, irresponsible being. These ghettos, the projected underside of Dear and Flusty's (2001) exclusive, interdictory spaces, are the ones whose class and cultural content necessitate that they be isolated in new global times.

Featured Elements

But these narrations of hopelessly lost black ghettos frequently have an even finer texture: many also chronicle an expedient cast of supporting characters and processes (particularly when longer, fuller narrations are offered). These, as crucial ingredients, give form and meaning to the narrations as Norman Fairclaugh's (1992) radiant and compelling "presences." The black family, a prominent featured and key element, is staged and choreographed to take a moral pummeling, albeit in presentations that frequently offer tragedy and irony. Perverse creations composing this family – "welfare mothers," "absent fathers," "hardened teens," and "on-the-dole parents" – are sites to unveil ghetto dysfunction at its most extreme. A textual strategy makes each a hybridized construct that blends an old cultural form – the aimless, poor black wanderer – with a more recent form – the low-income black huckster (see Wilson 2005). One poignant rendition of this family by writer Fred Reed (2002) is illustrative:

> We do not speak of . . . underclass [families]. We need to begin. Go into the homes [South Chicago] where in mid-afternoon a half-dozen men sit listlessly before the television, along streets where they sit for hours on stoops, doing nothing. There is nothing for them to do . . . the culture of the ghetto [and family] resists change . . . no academic urge is found in the ghetto, no entrepreneurial vitality, none of the traits that make for success in a techno-industrial society."

Most attention is typically given to the widely understood heads of families: fathers and mothers. To assault them is to unequivocally finger these families as leaderless and adrift. In the narrations, fathers and mothers are frequently made shiftless and transient: fathers unapologetic migrants from responsibility, mothers dreamers of a potentiality to leave families and wander. This portrayal of "black ghetto family" is distinctive. Mainstream portrayals of this family, to Robin Kelley (1997), once had these parents basically aware of societal norms but struggling (unsuccessfully) to keep families intact. Portrayals in the 1980s, for example, often rendered these families, in the words of columnist Thomas Sowell (1984), "increasingly removed from the culturally and socially normative." But in these current renditions, their awareness has disintegrated. In communication, as the social order of black ghettos and inner cities has broken down, so too has this black family's structure.

In this context, mothers are often illuminated as the centerpiece of this collapsed family. A prominent theme has them embittered and discordant but still exhibiting a shred of maternal instinct and decency. A projected personality, wrought from discursive bits and pieces of ghetto life – an incendiary mix of under-employment, lost husband, atavistic essence, and longing for better times – constitutes their essence. Revealed, for all to digest, is a hot-tempered, unevenly functioning woman whose crude maternal demeanor is the cultural glue that keeps these families intact. A devastating outcome is said to follow: a grossly ineffectual style of raising kids. Commentator Mancow Muller (2001) frequently conjures up and chastises this mother this way in the crudest of articulations: "These mothers are a problem, they're supposed to be raising kids, but more often are out there having a good time . . . the next generation is relying on them, and they're failing . . . what are they doing? Drinking, slumming, partying, not watching their kids . . . Yeah, there's caring here . . . but it's a strange kind of caring."

Sometimes, the staging extends to offering a caricatured intimacy of the subject, with the narrator provided a privileged, insider's glimpse into this mother's heart and soul. Mothers are most frequently placed in the home or the streets amid the turmoil and tumultuousness of child-raising and daily survival, where they become abruptly othered and rendered deficient. With "readers" peering in, cultural dysfunction becomes all-too-clear. As an illustration, *New York Times* writer Felicia Lee (in Wilson 2005) interviews subject Dona Williams on a Harlem street corner where the action is non-stop and frenetic. Here Williams struggles to survive and mentor already street-indoctrinated kids. The dilemma, to Lee, is that such women "never attend high school . . . [have] no marketable skills . . . [stay] home knitting and watching bootleg horror films as family members stop by . . . have eyes [that] are ancient." In this context, to Lee, kids not surprisingly turn elsewhere to meet their moral, social, and spiritual needs: peer groups, gangs, and buddies.

But amid these renditions, black youth are the featured element in these depictions of black ghettos. This youth, elaborately specified but grossly caricatured, is built around a central dilemma, the specter of cultural dysfunctionalism. These kids, as confused souls, are now easily seduced into playing rather than working or being civically engaging. Once dimly aware of mainstream norms and expectations, they have now ceased attempts to integrate into city social life: ghetto worlds preoccupy them. Parks and street corners, not homes or libraries, are their chosen social spaces. Exerting a raw physicality (playing basketball, roaming the streets), rather than cultivating themselves as good citizens, is the exercise that energizes them. Past successes in ravaging these kids (in discourses), strengthened especially in the Reagan 1980s, paved the way to now embellish this villain in stark symbolization.

In this context, these kids casually assimilate the haunting "underclass life." They are "core-less" beings (i.e. morally and ethically nebulous, and drawn to a rough and tumble street). Days spent pursuing the lures of fun and immediate excitement (street-corner hucksterism, selling drugs, malling, roaming the streets) animate them. The street, their favorite play space, is a cherished test site for their manhood and a place to enjoy the show. This space, an improvisational site for bizarre, sub-cultural encounters and engagements, becomes their emotive home as they spurn school, work, and home. "Streetwise" commentator Mancow Muller (2001), again, notes: "too many kids in the ghettos mindlessly follow the pack, " "become indoctrinated into the street life," "shun their families for the values and morals of the streets." This pull of the streets, in his narrations, seems beyond resistance. "Kids are pulled into this world," Muller comments, "because it's all around them . . . a gimme and let's hurt 'em world . . . wherever they go."

In this context, many of these portrayals temper this youth demonizing with notions of at least crudely feeling kids. These kids are frequently provided a shallow layer of civility for people to see through. These offerings make these kids superficially civilized, endowing with a vague comprehension of existing in writer Glen Loury's (1996) "drug infested, crime-ridden central cities." But as malfunctioning creatures, as Mancow Muller (2002) puts it, they "don't know how to change or even want to . . . they're having too much fun." These kids are thus projected to roam across inner cities bearing a vague semblance of civility, but, in what really matters, have their minds, souls, and aspirations buried in depravity. Here a generic inner city kid suffers from Dinesh D'Souza's (1999) "cultural dysfunctionalities in the black ghetto . . . that drives [these kids] to be hedonistic and present oriented."

At the same time, this construction of youth is commonly deepened through a deft ploy: marking bodies in presentation. Bodies, Wilson (2005) chronicles, are one widely used new instrument in post 1990 neoliberal discourses to unmask and make transparent "the real" of black ghetto youth. Narrating these bodies in detail charts a political anatomy of black subjects that allows narrativists to effectively condense assertions about the values, morals, and temperaments of youth and communities. Supposed ills of inner cities in this anatomical charting are made to powerfully appear on bodies – faces, eyes, legs, looks, mode of dress – that poignantly communicate. Bodies, to belle hooks (1993), are ideal narrative instruments: their inscribing is infrequently politicized in common politics and thus can be effective communicative devices. This treatment ultimately asks people to read bodies as a window onto their identity. So coding bodies, placed in convenient text, is a poignant way to communicate pathological kids.

Signifying black bodies in this representational strand proceeds in a process best described as heteroglossic (see Bakhtin 1981). Here, signifying always involves interaction between two forces: the absolute character of the signifiers and their situating. First, signifiers to work must reference a fixed system of meanings in common thought. But, second, these signifiers also communicate via the contingent spatial and temporal context of the signifying. Thus, marking these seized objects – black bodies – analytically dismembers and codes by references that everyone knows (e.g. being tattooed, transgressively glaring, and being physically imposing). At the same time, the relentless spatial and temporal "set-up" coding – placing these bodies on the stage of dark, decaying inner cities and in turbulent city times – fully grounds and "colorizes" the sense of feared and furious beings. This latter stage completes the circuitry of projecting the desired constellation of meanings: all dynamically coalesce into a unified, coherent representation.

A common bodying tactic used in these virulent narrations codes the faces of these kids as raw and atavistic. These kids, always grounded in foreboding streets and tumultuous city times, are expediently given the power to look but flagrantly reveal a now ruinous core: they can only glare. In one revealed, elaborately "placed" human instant, the essence of this new kid is supposedly unceremoniously unpacked: they can see the world but distort it and fail to comprehend it. In narratives, these are kids that "glare intently," "watch the streets behind scowls," "have the grimness of the streets etched on their faces," and "stare in intimidating pose at passersby" (see Norman 1993; *Chicago Tribune* 1994). These raw looks, a collapsed signifier, are a metonym for ominous and malfunctioning kids. A sight, melding with demeanor and setting, is made to reveal a blinded and problematic youth. Their eyes become, in such narrativist hands, points of seizure and clarifying in a "political anatomy" that forwards the offering of the "real" black ghetto youth.

But the most powerful signifier used in this "facial arsenal" is the ghetto scowl. This signifier, in the discursive setting of dark, unproductive ghettos in ominous global times, communicates discordant, angry, and incendiary people. Renditions of black ghetto life from commentator Mancow Muller (2001) are illustrative. Mancow speaks of "tough, aggressive punks . . . staring down anyone . . . kids on street corners . . . seeing everything that moves." These kids, to Muller, "glare at outsiders with anger and disgust." Muller's kids tenaciously resist being read: facial expressions seek to obstruct the "outsider's" gaze. This look of visual aggression shows a disdain for being objectively assessed: the light of truth about themselves had to be deflected. In this way, these kids are set up to revealingly spurn the most elementary act of human civility: simple observation. They are unable to stand simple contact with "outsiders", which communicates a central theme: an immersion in a cultural wilderness.

Speech of kids from a seemingly uncontrollable mouth is also frequently scripted by these narrativists to harden the impression of this projected youth. Spotlighted talk extends Reagan's syncratic imagery of confused and primitive speaking kids. This ploy, too, is time-tested: to Eric Cheyfitz (1981), conservative racial ideology commonly communicates the incompleteness of full humanity without eloquence and standard intellect. Offering brutish thought with conspicuously bad language displays part beast, a kind of monster. Illustrative of this, the Chicago Gang Research Project (2003), a widely accessed storehouse for gang data by the media, strategically "voices" a feared urban icon in Chicago and beyond: the Latin Kings. The voice of their local leader depicts brutish and primitive beings: "We here in the Motherland, extend our love worldwide . . . We must keep our destiny alive! Our Almighty Latin King Nation Manifesto and Constitution in our Heart & Soul. IF YOU DON'T ALREADY HAVE IT, GET IT, READ IT, STUDY IT, LIVE BY IT, IT'S KINGISM TO THE FULLEST!" This voicing goes on: "True Kingism is what we seek, what we demand! Learn what it means to live the life of Kingism!"

Seizure and scripting of African-American youth bodies in this rhetoric is anything but surprising. Such renderings of bodies, to Tommy Lott (1999), have a long history in the U.S. In the era of slavery, many narrativists focused on generic physique (huge brutish people), the early twentieth century on nimble, lithe bodies (minstrelsy), the middle twentieth century on fluid body parts (mobile urban people). Black bodies throughout, in the words of Cameron McCarthy (1995), have continuously been used as potent semiotic cargo. Current politicians, writers, and the media in these cities continue this process, but more profoundly tie these black bodies to something resonant: the sense of barren ghettos. These spaces in common imagining and thought had been established as zones of social and cultural otherness: their use in this way is adroit. In a spatialized communicating, these kids and others can offer all manner of excuse about their values and predicament. The "truth" of their character is revealed in something unmistakable: their talk, movement, and appearance set in garish spaces.

A final central subject in this ascendant strand of stigmatizing black ghettos within the new pattern of ghetto symbolization is a clear 1980s carryover, "drug-peddling gangs." Ghetto gangs, in much 1980s Reagan-era oratory, were grotesque objects that roamed and wreaked havoc on inner cities (see Collins 1996). Post-1990, in this virulent representational ensemble, they are an equally destructive group in devastated settings and now mired in worlds totally unto themselves. "There is no longer an outside world to so many of these kids," St. Louis Planner S. Plann (2004) notes, "only the stifling ghetto and the . . . reality of the gang." Their consciousness has changed: Once profound feelings of difference and alienation from the mainstream – creating anger and discord – have given way to what Plann called . . . "a full-fledged sub-cultural normalcy". . . creating "a norm of anger and discord." Ensnared in dysfunction, gang members know only their own and the gang's immediate needs. These villainous gangs, in this scripting, now duel with families for the hearts and minds of kids. Befitting the story-line, gangs are winning, families are losing.

This offering of garish gangs is typically nuanced via contextualization: they are vaguely linked to mentions of deindustrialization, post-industrial society, and pervasive racial discrimination. But the prominent storyline is gangs that are unremorsefully engaged in ongoing and destructive actions. Against this compelling and colorfully wicked storyline, the results are unsurprising: mentions of context and broader process fade into insignificance. Across neighborhoods, mobile kids move seamlessly across inner city terrain and tussle with corrective institutions – churches, Boys Clubs, legal aid societies, politicians – to define the social fabric. As city commentator Steve Nawojczyk (1997) notes, "street gangs are very fluid in nature . . . and dominate ghetto streets." Against them, to Nawojczyk, local institutions like tenant groups, block clubs, neighborhood watches, and churches are too often forced to recede to the twilight under the deadly spiral of crime and social disorganization. To Nawojczyk, gang activities "have spread to streets, schools, playgrounds, parks, and most frighteningly . . . to the world of common sense."

These depictions typically offer melodrama replete with rapid action and extreme states of confrontation. Riveting pathologies are made to sear these kids amid the spotlighting of a lurid cast of characters: unemployed men, angry kids, ex-cons, idle youth, and prisoners turned gang leaders. In the standard tale, gang members hail from dark, troubled, and wicked origins (bad families, bad neighborhoods). Upon membership, attempts to obtain group status and support lead to a rapid downward spiral of drug-selling and irrational violence. Now, the police can barely stem the tide. They confront kids who are often fearless, flaunt anti-societal values, and taunt the police. Moreover, for those that get arrested or locked-up, there is typically a confidence of being out soon and picking up where one left off. In the world of gangs, this means unfinished business – selling drugs, seizing new turf, recruiting new gang members, and involvement in the likes of prostitution rings. Nawojczyk recites this repetitiously rehearsed script:

> Young peripheral or associate gang members get their first exposure to the Gang culture through various aspects of the media – news shows,

> movies, videos, and even through the music of various artists. [These] glamorize the gang lifestyle. [for those that go] to prison, these youngsters become . . . indoctrinated into the world of real life gangbangers . . . Then, back to the streets these bangers go with more "knowledge" than ever could have been gained in the streets. When they are in prison, many gain rank or "juice" within their gang because they went to the "joint."

But this presentation of gangs is also nuanced in another way. At its heart is a narration of the gang through a time-tested group of metaphors: consumption and production. This kind of representing, to Anne Norton (1993), is common and is a powerful politics. These metaphors, she notes, are ominous signifiers in American thought. To take consumption first, diverse, potentially problem things are imagined to expand through this: foreign countries, territories, political systems, diseases, and armies. Consumption commonly invokes a sense of aggressiveness and expansive desires. Not surprisingly, sense of the U.S. being consumed by invading nations, internal cultures, waves of immigrants, and its own excesses populates American mythology. Production, similarly, is a powerful signifier to portray under-achievement and ineptitude. Through this metaphor, America's most marginalized populations have been symbolically ravaged (see Wilson and Grammenos 2000). This use of the two metaphors – a double-barreled signifying – ultimately assigns dangerous connotations to these gangs. Both point to an aggression and lethargy that asks the public to see these gangs as markers of community devastation. Let me elaborate.

On the production side, in the representing, gang members are in tatters. These kids have no desire to hold down "legitimate employment," only to illicitly sell contraband or not work at all. Legitimate employment is simply not part of their normative world. These deficient producers, to Indianapolis planner B. Braggs (2004), "are plagued by having no skills," "lack an entrepreneurial spirit," "too often fall into a problem lifestyle." Their main entrepreneurial undertaking, the quick and dirty selling of drugs, is lucrative, easy, and deliciously illegal. Civic and productive energies, to Braggs, no longer have a place in the mindset of these kids. Here is Braggs' "new reality of ghettos . . . hitting rock bottom." These kids, wounded by post-industrialism and now reveling in its dark side, reflect a "breakdown of community, family, and work – the heart and soul of civilized society."

On the consumption side, gangs have become commentator Mancow Muller's (2002) "circulating bunch of destructive kids." At the epicenter of the black ghetto fall is the ascendancy of this fallen, voraciously consuming youth. To Muller, these gangs "are out there pounding the pavement" and "looking for more turf." They "want [to occupy] more terrain," "confront anyone in their path," "are insolent and ready to be violent," and "will strike out at a moment's notice." "Like an army in gear," to Muller, "they thrive on action and are ready to go to war." Muller's gangs exude a primitive, black ghetto ethos that courses through their reasoning and thinking. Muller, like others here, talks about the difficulties of these kids' plight and circumstance and proclaims an open-mindedness, but in the next breath, happily reveals profligate kids needlessly eating up neighborhoods. These gangs, represented as ceaselessly devouring children, grope to understand themselves and the world that they find bewildering.

At this point, it must be reiterated that this virulent symbolic ensemble is but one major strand in this new regime of portraying black ghettos. It shares the spotlight with other multi-textured offerings as each repetitiously appears in diverse kinds of local "texts." But the influence of this strand is major: it appears trenchantly in local newspaper reporting, politician oratory, and planning pronouncements that feed ascendant neoliberal sensibilities. More flagrantly than the softer strands in this new regime of configuring or other kinds of accounts, it harshly offers fundamentally lost spaces and people. This ensemble, ultimately, offers something both new and not new: a full blown clinical disgust for completely decayed ghetto morals and social worlds that is staged via new and established metaphorical and linguistic tropes (see McCarthy 1995; Wilson 2005).

The new ambiguous ghetto-prison connection

A final emerging trend in these black ghettos must be discussed: the increasingly ambiguous ghetto-prison connection. Since the 2001 international recession, the seemingly calcified link between black ghettos and the prison-industrial complex as a containment complex, now widely chronicled (see Parenti 2000; Street 2003), has grown more tenuous. In a see-saw relationship, a social space once distantly linked to the functionality of prisons (before 1970) had this bond dramatically strengthen thereafter only to again weaken after the post-2000 recession. The first phase of this, the non-reliance on prison phase, was when these ghettos efficiently delimited, marked, and spatially isolated laborer African Americans without the need to draw on prisons and create a single "carceral" continuum. As William Tabb (1974) chronicles, black ghettos in U.S. cities before 1966 operated in a harsh societal era of separate and unequal. These ghettos, bolstered by the clout of societal belief and policy, efficiently stigmatized, constrained, and confined. Black ghettos then, paraphrasing Tabb, were America's most efficient mechanism for making, isolating, and using African-American labor yet devised.

But this efficient spatial apparatus to mark and isolate was weakened in the late 1960s. In the face of civil rights legislation and racial unrest that assaulted the logic and workings of this ghetto, change was in the air. Part and parcel of this, passage of local and national fair housing legislation, anti-redlining regulations, and more rigorous Realtor monitoring in more racially conscious U.S. cities threatened socioeconomic integration of urban fabrics. Suddenly, the functionality of the black ghetto was threatened. In short order, a response followed. Paraphrasing Loic Wacquant (2002), as the walls of ghettos shook and threatened to crumble, the walls of prisons were correspondingly extended, enlarged, and fortified. The result, between 1970 and 2000, was a surge in prison building of 400 percent, a 60 percent increase in rate of black incarceration, stepped-up policing of black ghettos, and the Justice Department's budget growth of 900 percent, that signaled the rise of Wacquant's (2002) single carceral continuum (Parenti 2000).

Yet by 2001, this ghetto-carceral connection had begun to unravel. At the core of this, a powerful new recession, major federal withdrawal of resources from states and cities, and massive funneling of dollars to the military and the Iraqi War, has meant budget crises across U.S. cities and states. Like the mid 1970s, when many rust belt cities faced fiscal default, financial standing of cities and states became ominously unstable. Many mayors and governors, therefore, woke up to recognize the massive expense of building prisons and bloated incarceration. Now, as I discuss, politicians are rethinking and beginning to scale back the rise of the prison industrial complex. Its sheer cost in an era of perverse fiscal priorities and economic crisis makes the black ghetto-prison nexus a sticky and ambiguous proposition. This possible change in ghetto-isolation dynamics – across the entirety of rust belt cities – has potentially profound reverberations that are discussed in this section.

The 1990s Backdrop

Black incarceration rates and prison spending, following a pattern established in the early 1970s, continued to escalate in the 1990s. The extending and enlarging of prison walls that proved so adept at marking African Americans and putting them in a circuit of social and economic marginality was in full swing (see Wideman 1995). Statistics across the rust belt reflect this. By the late 1990s in Iowa, prisons had grown beyond capacity. Iowa's nine prisons, with a capacity of 6,772 inmates, exceeded 8,200 by 2000 (see *Daily Iowan* 2005). Similarly, Iowa courts sent an average of 300 convicted offenders to the state's nine prisons each month in the 1990s. In Illinois, the prison population spiraled out of control, with inmate numbers increasing from less than 7,500 to over 43,000 between 1970 and 2000 (Street 2003). In 2002, more than 7,600 of these people came from just six of Chicago's 66 zip codes, five on the city's south side and one on the west side (Street 2003). During this period, the number of prisons in Illinois rose from 7 to 27.

Yet, overcrowded prisons in Minnesota and Indiana were even worse. In Minnesota, police crackdowns and stiffer sentencing resulted in the state's ten correctional facilities having fewer than 75 empty beds. In an emergency measure in 2003, Governor Tim Pawlenty asked the legislature to approve borrowing $95 million to expand two state prisons. In Indiana, where black Indianapolis inmates account for more than 30 percent of the prison population, the state between 1996 and 2000 increased its spending on prisons by 71 percent (see Gainsborough and Maves 2004). But the cost, more than six times the rate of inflation at this time, failed to adequately address the situation. In 2003, to state officials, crowding in Indiana's prisons was at its worst level in four decades. Indiana's two central prisons, intended to hold 16,000 inmates, contained nearly 23,000 (see Buck 2004).

Crackdown in black ghettos across the rust belt fed this expanding carceral system in the 1990s. Stepped-up policing, zero-tolerance policies, and stiffer drug laws and sentencing came to permeate these inner cities guided by neoliberal and revanchist sensibilities (Goldstein 2001). The results of the crackdown show in national statistics: a rising penal population became much less white and more black. Non-Hispanic whites accounted for 42 percent of state prison inmates in 1979 but less than 33 percent by 1999 (Street 2003). Blacks, 12.3 percent of the U.S. population in 2000, comprised roughly 51 percent of the nearly 2 million incarcerated people across America. Between 1980 and 2000, the number of black men in jail or prison increased 500 percent, resulting in more black males behind bars than enrolled in U.S. colleges of universities (Street 2003). In a typical day in the 1990s, 30 percent of African-American males aged 20 to 29 were under correctional supervision, either in prison or on probation or parole (Buck 2004). In 2000, the incarceration rate for African Americans was 1,815 per 100,000 compared to 235 per 100,000 for American whites.

But ghetto crackdown and incarceration were only parts of this expanding black ghetto-prison system nexus. The machineries of stigma, hard at work, powerfully marked prisoners as felons upon release, which produced a massive army of stigmatized and spatially manageable "ex-offenders." The prison walls, in this sense, extended far beyond the prison. Each year in the 1990s, more than 600,000 people were released from state and federal prisons, feeding a swelling army of ex-offenders bearing what the *Economist* called "the stigma that never fades" (see Street 2003). The result, as Loic Wacquant (2002a) reports, was devastating: many African-American ex-offenders were forced back to black ghettos, where confined social spaces and barriers to finding decent employment perpetuated the closed cycle of poverty and marginality. The essence of socio-spatial confinement was unrelenting (e.g. many were forced to disclose their time incarcerated to gain work, resulting in being bypassed for the dwindling supply of decent paying jobs, denied rental or purchase of housing, and prevented from obtaining a credit card). If ex-offenders did not find their way back to prison, they often eked out lives at the bottom rung of ghetto communities.

By the late 1990s, then, the black ghetto as a mechanism of naked exclusion had been fortified with key help from the state's penetrative penal arm. The prison, elevated to the rank of central machine for race making, routinely spun out derogated and afflicted people who were to assume their rightful place in the color-coded city: the black ghetto. The post-Keynesian racialized economy would efficiently mobilize them, notably in the burgeoning residual and fast-food sector, all the while their bodily presence would be meticulously restricted. The prison, now, reached deeper into society and the city as a sanctioned, growing societal institution. Shedding most pretense at being a site to rehabilitate and heal, it became avid producer and regulator of this societally scorned population. In this era, deepened black poverty and the societal cultivation of an anti-black animus became its visible productions.

The Present

But since 2001 this rush to incarcerate and build prisons has become increasingly problematic. Re-thinking this strategy has followed from the one force that could substantially impinge on this prison functionality: a debilitating economic recession and the reality of financially plagued cities and states. The recession officially began in March of 2001 with more than two consecutive quarters of decreases in gross domestic product. It has continued in the guise of a mini-recovery that is labeled even by conservatives as "anemic" and "all but non-existent" (see Irons 2003). In the process, federal revenue has declined to record levels amid unprecedented deficits (Irons 2003). In 2002, the $455 billion deficit had increased to six percent of gross domestic product, a historically unprecedented amount. President Bush's 2002 and 2004 cutting of federal taxes and government spending have, not surprisingly, exacerbated this condition.

As early as mid 2000, the impact of the economic downturn on states was obvious. Fueled by growing joblessness, thirty states projected budget gaps totaling $40 billion in 2000 (see Pierce 2003). For example, New York, New Jersey, and Arizona projected deficits of 13, 21, and 17 percent, respectively, which materialized by 2002 (Pierce 2003). At the same time, many cities and states saw their bond ratings downgraded as they struggled with funding cuts, making it harder to secure financing. Conditions in cities deteriorated from there. As Bush capped the federal contribution of Medicade and demanded that state and city governments absorb the costs of homeland security (e.g. hire more firefighters and police), finances grew worse. In Cleveland, Detroit, Chicago, St. Louis, Pittsburgh, and Philadelphia, government employment was cut by more than 20 percent to deal with dwindling revenues. In 2003, the National League of Cities Annual Survey found that more than four in five cities (81 percent) were less able to meet financial needs compared to the previous year. In these cities, spending increases outpaced revenue increases by 3 to 1 percent.

In this environment, many state officials have become vocal about the fiscal implications of sustaining the prison industrial complex. Ironically, these same officials, in previous years, typically embraced the idea of more prison building. For example, one-time prison-construction advocate Mike Lybyer, Missouri Senate Appropriations Chairperson, now laments the Department of Corrections (DOC) half-billion dollar budget that now squeezes higher education. He notes that the university system would "have a lot more money if you could tell us how to keep from building more prisons in this state." President of the University of Missouri Board of Curators Malakin Horner concurs, noting to the press: "I don't think anyone thought corrections would overshadow our mission of education." Reducing prison construction, to Senate chairperson Lybyer, "could be one of the best things you do for higher education" (in Taylor 2000).

In Indiana, where taxpayers now pay almost $21,000 a year to house and feed each of the over 1,200 state inmates, the talk is similar. Governor O'Bannon, facing an $800 million deficit, now talks routinely of "reforming the monstrous state penal tiger." In a few short years, O'Bannon has gone from extolling the disciplining effects of prisons to invoking them as economic liabilities. Even to conservative Senate President Pro Tem Bob Garton, a re-thinking is in order: "it's time to look at the sentences because that is causing a lot of your prison buildup" (in Center for Juvenile and Criminal Justice 2002). Garton, not surprisingly, routinely talks about the need to have a fiscally responsible government, but, very surprisingly, notes that this should include reining in prison building and prison spending.

This talk of halting expensive prison building in Indiana now mixes with the neoliberal tenet of privatization to bolster the latest prison "reform:" privatize the existing prison network. Indiana's new director of prisons now suggests that any new prison be built and operated by private companies. Department of Correction Commissioner David Donahue, a former vice president of private company U.S. Corrections Corporation, pronounces this amid expectations of the state prison population climbing above 25,000. The sheer expense of any new such construction in this era of economic malaise, Donahue argues, should be passed to companies, not taxpayers (see *Associated Press and Local Wire* 2005). Donahue, of course, also trumpets this move as "efficient" and "streamlining of prison operations." As of summer 2005, the move was already underway: the state had begun accepting bids for privatizing health care provision, nurse care provision, and food preparation and distribution in state prisons.

Commentators and the media have joined the chorus of dissenters. Freelance Wire Services (2004) calls the California Department of Corrections "a dysfunctional agency . . . with layer after layer of scandal and deceit, and a stench that only gets stronger as each layer is removed." The supposed dilemma? A growing and out-of-control prison complex and bureaucracy. Here, Freelance Wire Services charges, prison officials grow rich and hire with impunity as funds flow their way, in this case costing the state $100 million. And almost not to be believed, conservative writer Fox Butterfield, an avid supporter of unfettered prison building in the 1990s, asks the poignant question: "if the crime rate keeps falling, why is the number of inmates in prisons and jails around the nation going up?" Butterfield now speaks about an irresponsible overbuilding of prisons in an era of tight fiscal times.

Unanticipated actions have followed. In 2004, 567 Lexington, Kentucky inmates were abruptly ordered released by Governor Paul Patton to help reduce a $500 million budget deficit. All were nonviolent offenders who had previously been convicted in the state's recent "get tough on drugs" crusade. Iowa, facing a record $120 million deficit, laid off 300 prison guards as a cost-cutting measure. In Oklahoma, conservative Governor Frank Keating made the Pardon & Parole Board free 1,000 nonviolent inmates after adding over 1,000 new inmates each year between 1998 and 2001. He claimed that the state's budget crisis made this a necessity. In Virginia Beach, Commonwealth Attorney Harvey Bryant, the local prosocuter, declared that state cutbacks to his office would prevent him from prosecuting most of the 2,200 misdemeanor domestic violence cases he gets each year. No opposition has followed.

At the same time, policies and laws are now beginning to be affected. Michigan's legislature, faced with a record deficit and momentous prison overcrowding, voted to repeal the state's strict mandatory minimum sentencing laws for drug crimes. Previously, these same crimes led to some getting life sentences for possession of cocaine or heroin (see Gruley 2001). In Kansas, the Sentencing Commission successfully authored a new policy where people arrested for drug possession, with no record of prior arrests for violent crimes or drug trafficking, be placed in mandatory treatment instead of prison. About 5,000 of Kansas's 9,000 inmates would be affected, the cost per person going from $21,000 per year (in prison) to $2,500 (in drug treatment). As concern grows about the sheer economic realities of the prison-industrial complex, and the national economic malaise persists, more of this should occur.

Black ghettos, now, are beginning to be choked off from a major apparatus that has reinforced their character and function even while the trend is in its infancy and how it will end up is unknown. The dilemma, to growth elites, is the role that the ghetto-prison carceral system plays, seamlessly constituting the stigmatized persons to be shunned and isolated. But these ghettos, even with the withering of their supportive prison industrial complex, are not in danger of disappearing anytime soon. Put simply, they are too important in this current rust belt restructuring: they help engineer the current accumulation lifeblood of these cities, i.e. renewed gentrification, commodified ethnic spaces (e.g. Chinatowns, Greektowns, "authentic" Mexican enclaves), high-brow public spaces, and CBD transformation. It follows that if this prison apparatus evaporates entirely as a ghetto-nurturing mechanism, which is not likely, other tools would be found to replace it. The size of prisons can shrink, but the role of black ghettos as a warehousing instrument cannot.

If this ghetto-prison nexus continues to unravel, two repercussions are likely. First, policing of black ghettos and public spaces will probably intensify. Denied the unfettered use of prisons to stash, isolate, and mark this population, such a compensatory response would be anything but surprising. That would mean more drug raids, more curfew and no standing ordinances, more intense surveillance of streets and public spaces, and more targeted auto stoppings in cities. The heightened "surveillance city" of Los Angeles, which Davis (1990) so eloquently describes, could become the norm across rust belt cities. Indeed, this appears to be happening already, where such stepped-up tactics in Chicago's South Side, Cleveland's East Side, and Indy's Eastside are being widely reported. In these cities, there is a supposed growing youth problem in a new era of lurking terrorists and anti-American malcontents. In this context, local courts here continue to sanction the police's use of race as an index of increased risk of criminality while legal scholars have endorsed it as a rational adaptation to persistent crime (see Wacquant 2002).

In a second potential repercussion, the machines of local stigma production may heat-up to more deeply impugn and marginalize black ghetto residents. The already shrill revanchist rhetoric chronicled about black ghettos and their populations would intensify, becoming even more hyper-virulent. Numerous local columnists, pundits, and politicians, whose work-related constraints and political predilections typically align them with dominant growth visions, would continue to be key constructionists. Stigma, as discussed, has been and is currently a key ingredient to controlling and confining this population; managing people in the neoliberal era is as much done by producing stigma as construction and implementation of rules and regulations (see Weber 2002; Keil 2002). Ultimately, the use of representation to make a problem identity, on top of a stepped-up monitoring and policing, would help keep black ghettos workable for a hyper profit-searching real-estate capital.

5 Recent sustaining – Bush policy effects

INTRODUCTION

Key issues around "the national influence" still remain unexplored. What are the effects of new federal policies on these ghettos? To what degree do they promote this marginalization and deprivation? And how are we to understand these influences? George Bush's initiatives have been plentiful and ambitious as neoliberal discursive and material projects. These policies, chronicled in this chapter, permeate rust belt cities and help drive this third wave of black ghetto marginalizing. These initiatives, notably enhanced Workfare, No Child Left Behind, and Faith-Based Resource Provision, reflect the new neoliberal times as federal and local initiatives coalesce to push new social regimes, educational policy, and economic development. These ghetto spaces and their populations, perhaps unsurprisingly, are posited as under-achieving and civically unproductive elements in new global times, at a concrete level something resembling what Pile, Brook and Mooney (1999) describe as "urban jungles" infested by dangerous strangers.

To understand these influences, I suggest that Bush's federal policies can be seen as inseparable from the local global trope that afflicts rust belt black ghettos. My point is simple but important: This package of rules and resources, originating at the federal level, becomes influential and potent as elements grounded and plugged into circuitries of local affairs. Borrowing Massey's (1999) geologic metaphor, these policies deposit "ideological and material sediment" in these cities which helps comprise local frames of understanding, institutional configurations, and bases of resources. Federal programs and policies, received and embedded in places, are activated by human intent to push the growth machine's agenda of making spaces, people, and the new entrepreneurial city. A mélange of federal programs, thus, become intensely localized things. For this reason, I reject the causative distinction between "local" versus "federal" effects on these cities and poor neighborhoods, even while it is useful to analytically pry them apart to understand their sources and origins.

This said, I suggest that these federal initiatives, like their local counterparts, are anything but random creations. In this case, they have been responses to the broader national totality (i.e. they can be seen as a kind of "institutional fix" onto a post-1995 moribund national economy mired in underconsumption and underproduction) (Sherman 2002). In broad brush-strokes, the dot com propped-up U.S. economy of the 1990s, struggling for years, finally fell into deep recession after 2000 fueled by this sectorial collapse (Brenner and Theodore 2002). After 2000, eleven consecutive quarters of decreases in gross domestic product proved devastating, withering jobs, manufacturing, and investment (see Street 2003). The U.S., once the world's mass-production Fordist King and more recently a strong post-Fordist force, was unsettlingly moribund. Yearly, trade deficits since 2000 have averaged $200 billion that extend to most world countries (including a $170 billion deficit with China) (Street 2003). Since 2000, the receding U.S. economy has lost more than 3 million manufacturing jobs – more than 10 percent of its 1999 base – with no end in sight (Stettner and Allegretto 2005).

But capital has continued to try to mobilize all territories and places as forces of production and consumption. Bush, pursuing yet another round of institutional fix, has moved even more decisively to re-configure the national institutional base, centered on re-defining the national state's already reduced social welfare role to use resources more entrepreneurially and "efficiently." With economic crisis crystallizing, more resources are to regenerate the ailing economy (i.e. subsidize new business investment, discipline workers to be more productive, more strictly regulate public schools to infuse curricula with technical and "moral" content). The resultant package of programs – modified Workfare, Faith-Based Resource Provision, No Child Left Behind, Block Grant deepening – has been a coordinated effort to enhance worker productivity, create a more pro-business regulatory climate in places, and promote "growth-first" social landscapes. The once dominant Keynesian political configuration, even more abrasively than before, has been pillaged.

A key analytical object of this Bush institutional fix, perhaps unsurprisingly, has been the global trope. This rhetorical formation, previously chronicled as an influential narrative in rust belt cities, thus has a parallel, inseparable "partner" at the national scale. The global trope therefore proceeds simultaneously at the "the ground level" and from above as mutually constituting discursive formations: each emboldens and shapes the other in a seamless dialectic that connects and melds scales and rhetorics. A sense of new global times and offering of policies, continuously invoked by President Bush and policy advisors, infuses local political climates with meaning and content which invigorates the national rhetoric and policy. Most poignantly for us, as Bush (2002) speaks and institutionalizes into policy the notion of "we all must now be more productive and responsible . . . from the farmer to the urban worker and dweller . . . for it is a new global day and nothing else will do," rust belt cities become "shot through" by a reservoir of ideological and material content.

A brief caveat to this is that national global rhetoric is simultaneously supportive and at odds, at an important level, with the local global rhetoric. On the one hand, the national rhetoric reinforces the local articulation of a new, undeniable economic reality and what must be done. Both formations speak notably of a compressed economic earth that has created footloose producers and the need for places and populations to be more economically productive. But the casting of the black poor is somewhat different. The national rhetoric typically identifies the black poor as potentially salvageable targets in need of cultural and social rehabilitation, the local rhetoric more often virulently demonizes and dismisses them. The first identifies a supposed unproductive people who need re-molding for the civic and national good, the second often modulates this with a casting of an all-but-hopelessly lost population. The specifics of these rhetorics may be surprising: it defies notions that the local is always the terrain of greatest sensitivity to social reproduction on all issues at all times (see Dear and Clark 1984).

However, this differential treatment of poor black communities and populations is explainable. In the local of rust belt cities, capital more than ever (in the era of gentrification and downtown transformation as core urban policy and accumulation strategy) needs to be signaled that black ghettos will be assiduously controlled and managed. Growth machines seek to constitute a tangibly new urban form, one that is attractive for new investment and new jobs that will enhance real-estate capital. In Chicago, Milwaukee, Cleveland, and St. Louis, businesses, builders, developers, and investors power local real-estate profitability and city revenues: their perceptions and predilections are widely tended to. With the era of hyper-real estate accumulation comes the pervasive fear that such accumulation can cease. No such immediacy exists at the national scale. Bush's agenda, less concrete, is to renew general economic health across society by providing the resources and conditions that power profit for businesses (particularly for big business). At this level, Bush and his people have the luxury of articulating abstract and emotive notions of unifying all citizens, even as the rhetoric communicates an intrinsic racial and class unequalness and the need to re-configure or punish unproductive people. That said, we may proceed.

BUSH URBAN POLICY

First, a brief sketch of Bush urban policy. Bush offers a welter of urban-based initiatives leading with the rhetoric that little policy should be explicitly urban. This rhetoric shores up his neoliberal and non-urban political base, and strikes implicitly at the social welfare apparatus. Bush's repetitious message is that policy should lift all boats rather than focusing on lifting an urban boat or any other "interest-group" boat. In this rhetoric, Bush policy is not about "urban giveaways" and playing politics with powerful constituencies, but about making America more competitive and compelling in new global times. The world has changed, in Bush articulation, and policy should aid America's ambition to find its rightful economic and political place in this.

But Bush's anti-urban rhetoric is not new: it involves a deepening along a historical continuum. His administration fits an approximate 20 year trajectory of an increasingly neoliberalized federal government that peripheralizes city needs. With the neoliberal experiment sticking, and Reagan effectively offering cities as a kind of profligate client state (with Bush I and Clinton adhering), urban needs have been marginalized (see Burnier and Descuter 1992). During Fordist–Keynesian times, presidents routinely visited cities and talked about their needs and concerns. The uniqueness of cities as places rooted in production with supporting low- and moderate-income assemblages of people was instilled in common thought. But the neoliberal project after 1980 dislodged this sense of distinctiveness from common thought. By the time of Bush II, this "urban exceptionalism" had become a distant memory. The word city now rarely enters his state of the union addresses, which has helped marginalize in mainstream discourse growing problems in this terrain, notably poverty, underemployment, and hunger.

Yet the reality of pursuing this latest round of institutional fix necessitates precisely what Bush admonishes: government involvement in cities. For regenerating the national economy also means the structuring of economic processes and social relations in a key spot – cities. To enhance general worker productivity and pro-business regulatory climates, this strategic place of production must be assiduously managed. In this context, Bush's new programs intermix neoliberalist ideas and religious principles with proclamations about new global times to privilege two institutions: churches and business. Both, positioned to more deeply manage and regulate use of land, spaces, and social conduct, are free to impose their disciplining sense of morals and judgments in the realm of work relations and everyday life. Times have changed, to Bush, and so too must social worlds, economic ways, and patterns of public-sector involvement.

But these programs, as will now be documented, profoundly afflict in their own distinctive ways as they help propel the third wave of black ghetto marginalization. Faith-based efforts call out and excoriate a supposed culture of the poor, Workfare punishes the poor for their plight as it mobilizes them for servitude in the new fast-food economy, and No Child Left Behind fingers and expels "underachieving, problem youth" from public schools. Moreover, unanticipated repercussions from these programs deepen the production of deprivation (e.g. growing numbers of the black poor are excluded from life-sustaining welfare, expanding numbers of teachers flee public schools given extraordinarily difficult working conditions, and the needy get stilted advice from religious counselors advising on family planning, social problems, employment, and child rearing). As rust belt growth machines toil to make urban space and social milieus more entrepreneurially taut, these programs have proven expedient and handy.

FAITH-BASED INTERVENTIONS

One centerpiece of Bush's urban policy now embedded within rust belt cities – "faith-based initiatives" – is advertised to shape a more productive and culturally unifying citizenry (i.e. locally and nationally civic, contributory beings) in a global era. This policy thrust has meant, in Bush's words, unleashing "armies of compassion" to help those with "troubled and counter-productive" values and lives. Benevolent and professional volunteers, in the rhetoric, will rally to help these people propelled by a powerful force: religious and moral rectitude. In Bush's (2002) words,

> federal policy should reject the failed formula of towering, distant bureaucracies . . . [and find] these quiet heroes . . . usually on shoestring budgets . . . that heal our nation's ills one heart and one act of kindness at a time. We will focus on expanding the role of social services of faith-based and other community-serving groups that have traditionally been distant from government.

But this initiative has roots in a curious history that must be recounted. Bush, as Texas Governor, hired former Watergate burglar Charles Colson to initiate his Prison Fellowship Program in two prisons (part of Bush's Inner Change Freedom Initiative). The initiative, straightforward, barraged inmates with an intense dose of bible-centered evangelizing, prayer, and religious counseling to instill proper values. Within weeks of the program's initiation, Bush hailed it as already producing positive results. He received key support a few months later, from the widely-publicized review of the program by the University of Pennsylvania's Center for Research On Religion and Urban Civil Society (CRUCS) (2002). The findings were sensational if predictable: program graduates were being re-arrested at dramatically lower rates. The Center concluded that religion was the ideal instrument to culturally engineer and normalize criminals. The days of excluding religion from social service provision, to CRUCS, should be over. Next, the results were proclaimed in a widely publicized 2000 photo op (led by Colson and Bush) at the White House and a triumphal press release by Tom Delay (Kleiman 2003).

Yet all is not so simple: close scrutiny of the report casts a shadow on the experiment's veracity, revealing trumped up results and a dubious program. A facile trick made the Prison Fellowship Program appear a success, counting the winners and ignoring the losers. Kleiman (2003) called this Bush tactic "selection bias," a.k.a., "creaming," critic Anne Piehl termed it "cooking the books." Of the 177 prisoners in the program, Colson ignored the 102 participants who dropped out, were kicked out, or were paroled and did not finish. The sub-sample of 75 who completed the "training" became the study's sample, ultimately faring better (i.e. getting jobs) than a control group. But when the sample is expanded to the original 177, compared to the study's control group, the group did worse. These grim facts, stuck in at the study's end after pages of discussion on the excellent results from the group of 75, deceive. Yet the White House was not put off: they continued to work off a selective press release that discussed only the group of 75 (which continues today).

In this context, Bush aggressively proposed his faith-based agenda that met initial resistance in Congress, which feared an intermixing of church and state. But this was overcome with a persistent rhetoric that relentlessly attacked two time-tested societal villains: "the misguided welfare state" and "a hopelessly bloated government." As the White House Office of Faith-Based Institutions (WHOFBI) (2003) noted,

> President George W. Bush's Faith-Based and Community Initiative represents a fresh start and bold new approach to government's role in helping those in need . . . Too often the government has ignored or impeded the efforts of faith-based and community organizations . . . Their compassionate efforts to improve their communities have been needlessly and improperly inhibited by bureaucratic red tape and restrictions placed on funding.

In this setting, Bush successfully issued Executive Order 132.79 in 2002 that made faith-based institutions eligible for HUD funds (see Department of Housing and Urban Development 2002). With this legislation, faith-based institutions that provided social services could be funded through federal and state community development block grants (CDBG), Home Investment Partnerships, Emergency Shelter Grants, and Housing Opportunities For Persons With AIDS (HOPWA). While faith-based institutions were to be funded and evaluated on merit and performance, their religious core could be exempt from scrutiny (i.e. they could retain their independence of governance and expressions of religious beliefs). They thus could constitute their boards on a religious basis, display religious symbols and icons, and hire religiously compatible employees. Religious activities – "worship and religious instruction" – were permitted if technically voluntary for program participants. Within two months of the Executive Order, HUD was advertising to sub-contract with 250 faith-based community institutions across America in its Reaching the Dream Initiative.

Faith-Based Initiatives Embedding in Rust Belt Cities

Faith-based initiatives quickly became institutionalized into the fabric of rust belt cities. But why the immediate adoption in typically heavily democratic cities? First and foremost, many growth machine actors (particularly mayors and council people) viewed the program as supportive in the drive to restructure and re-entrepreneurialize their cities. As Philadelphia Mayor John Street noted, "the faith based initiative is an old idea whose time has come . . . I believe it can make a real difference in our city" (in Maharoj 2001). At the same time, sheer pragmatics was at work: As writer Paul Street (2003) notes, clinging to beliefs of perceived best ways to help cities and the indigent is one thing, getting tangible resources is another. Like many city leaders had always done, to Street, practicality won out over any deeply felt counter-belief. Finally, as Peck (2001) describes in the recent embracing of Workfare, here was something new and advertised as combating an unpopular notion: inept and wasteful government spending. Faith-based outreach was packaged as something appealing in cleaning up both "the welfare mess" and a hopelessly bureaucratized government.

In Cleveland, faith-based resource provision has been embraced by Mayor Campbell. To begin with, the city's social service sector has been ravaged since 1980 by federal funding cuts with the sustained dismantling of the welfare state. By 2003, this local sector was approximately one-half the size of its former self in 1980 (Rentgen 2004). In this setting, churches could conceivably aid this sliding-into-oblivion sector. But a parallel force also powered this adoption: the desire by Campbell and the local growth machine to mobilize Cleveland's monstrous East Side black ghetto, which was to produce new, hard-working laborers in the city's ascendant service economy (both low-wage and high-wage sectors). These ghettos, paraphrasing Mayor Campbell (2003, 2004), needed to contribute to the "New Civic Cleveland", which church values and doctrine could supposedly assist (see Campbell 2003, 2004). To Campbell and others, times had changed. Cleveland, from its ghettos to its innocuous cul-de-sacs, was to be disciplined and brought into the new reality of global times, and churches were identified as one key facilitator.

In this context, the city and county almost immediately centered religious institutions as "the new noble help for social problems" (Jenks 2004). To Program Officer Marcia Egbert (in Absey, Darmstadler et al. 2004) of the George Gund Foundation, "faith-based organizations often have long-standing roots in communities and records of service." "They," to Egbert, "are a resource worthy of support." "If we ever need the Church, we need it now," said Ralph Johnson, general manager of local work training for the City of Cleveland. Opposition to date has been minimal. To local observer K. Williams (2004), "this faith-based thing [in Cleveland] has been so strongly sold, so widely touted, few have dared to seriously question it." "Even as federal requirements and levels of expertise have dropped in so many areas," Williams continues, there's been little controversy. "If you value your political career, even in this heavily democratic city, you go along with it."

Faith-based institutions in Cleveland now run food pantries, homeless shelters, drugs counseling centers, and youth mentoring programs (an estimated 35 faith-based organizations receive federal funds). The traditional bulwark for this provision, government agencies, have shrunk dramatically (Williams 2004). But the greatest amount of block grant funds is used to train, teach, and support the 5,100 black welfare recipients who must find work under Workfare. Welfare recipients are fast-tracked to work mainly in local fast-food restaurants and public maintenance positions for wages that hover around $5.30 per hour (Williams 2004). In Cleveland, as elsewhere, welfare recipients are limited to 60 months of welfare benefits in their lifetime and must work or train for 30 hours a week to keep benefits. This support is often done at centers that mix volunteers and paid workers (pay accruing to workers based on ability of institutions to obtain local or federal funds).

In Philadelphia, the story is much the same. The faith-based effort has been embraced by Mayor John Street and the City Council. Philadelphia experienced more than $40 million in federal block grant and social service cutbacks in the 1990s due to the national dismantling of federal programs. In this context, the city's social welfare network dramatically shrunk. Churches, again, have been envisioned as the new hope to step into the breach, Mayor Street's "innovative approach . . . [with a] long history . . . from the Quaker abolitionist movement to the Reverend Leon Sullivan's Opportunities Industrialization" (in Maharoj 2001). But again, another equally important force compels this adoption: the desire by Street and the growth machine to entrepreneurially sharpen the city's Fordist form. North Philadelphia's massive black ghettos, in this vision, are to be both insolated and re-engineered with new civic and work values and contribute cheap but vital labor to the city's new service economy. These ghettos, like Cleveland, are to stay where they are, become plugged into the circuitry of the "New Economic Philly," and, paraphrasing Street (in Maharaj and Bullock 2003), wear the church's values that are deemed ideal for renewing civically under-engaged populations.

In this context, Mayor Street has embraced this initiative and in 2001 established the Mayor's Office of Faith-Based Initiatives (MOFI). Street, once a radical political activist, now extols this program as "a key initiative in a long history of city faith-based activism." George Bush has noticed, hailing the Street Administration "a paradigm for partnership between government and religious volunteer organizations" (in *American Atheist* 2001). A black man, cast as converted from radicalism to the "truth" of Republican politics, has been an ideal Bush icon. Street, now a devout Seventh-Day Adventist, has raised more than $150 million to energize the undertaking. These organizations, as in Cleveland, increasingly supplant government offices across the city with Christian outreach entities. They provide food for the hungry, supervision to youth, shelter for the homeless, marital counseling and family planning information to people, and counsel drug abusers.

Street's support structure in this endeavor has been immense. He has been counseled by fellow Philadelphians John DiIulio (Head of Bush's Office On Faith-Based and Community Initiatives and then professor at Penn) and Reverend Herb Lusk (a Baptist Minister who to the *Philadelphia Daily News* is the White House's favorite inner city pastor). Both are close to Street and provide technical support in informal discussions and oratory in the local media. Moreover, both local dailies, the *Daily News* and the *Inquirer*, are also supportive in editorials and stories written. Not stopping there, Street, with Miami Mayor Manual Diaz, launched a new initiative in 2004: the Mayor's Center for Faith-Based and Community Initiatives (see U.S.

Conference of Mayors 2004). The Center, a national outreach initiative, seeks to "inform, educate, and train mayors, city-designated faith-based liaisons, and other public servants on how to best engage the faith community for more effective partnerships and service."

But all is not as simple as these political operatives forcefully suggest. Much evidence shows that these faith-based programs have evolved as aggressive political blocs. Debunked, most centrally, is the myth of these faith-based providers as passive, selfless helpers of compassion. Clergies across rust belt cities have increasingly organized to become powerful political as well as human-services forces, forming institutional blocs to be potent political voices that control tens of thousands of votes and potential volunteers for political campaigns. Their quest to increase their base of power and influence is often flagrant. For example, Philadelphia's Archdiocese and Philadelphia Baptist Association have allied with the City's Faith-Based Partnership Office. They now meld as a unified political voice to support and lobby for educational vouchers, No Child Left Behind, and Workfare across Philadelphia. Now, local Democratic and Republican parties recognize the power of the faith-based coalition, and seek their advice and support on such far-flung issues as community development, housing improvement, and school policy.

In this context, these city governments often court and reward key faith-based institutions with jobs and public service contracts. Two of Philadelphia's prominent faith-based groups, Reverend Herb Lusk's Baptist Ministry and Wilson Goode's Baptist Church, have continuously received funds to train welfare workers, conduct after school programs, and provide drug counseling (O'Hare 2004). Each staffs their faith-based workforce through a mix of voluntarism and government subsidy that perpetuates their privileged role in the service provision realm. And their power now extends way beyond the Philadelphia city limits. Lusk spoke at the Democratic GOP National convention in 2003 and hosted lavish party events during the convention week. A long-time supporter of George Bush, Lusk's programs have now received more than $1 million in grants (Common Dreams News Wire 2006).

At the same time, evidence suggests that this faith-based thrust poses new kinds of problems to clients. The dilemma is the use of volunteers, use of workers exempted from state and local regulations, and the tendency of many workers to extol if not proselytize religious beliefs. Now, authoritative voices of social service provision increasingly orate and work through religious tenets, reduced regulations in service provision, and cost-cutting of general operations. These voices poignantly mirror and push neoliberal beliefs: privatize public resources, reduce wages of workers (or use volunteers) to make organizations more efficient, re-make the poor with spiritual and entrepreneurial values, and kill off welfare politics. In this context, the inadequate providing of human services to the poor (documented shortly) is anything but surprising, with provision often incomplete, religious-driven, and makeshift. Treating clients as spiritual and cultural basket-cases, to paraphrase Chicago organizer K. Williams (2004), not surprisingly leads to basket-case provision.

For example, Philadelphia's ascendant clergyocracy is a closely knit set of organizations that at best incompletely addresses the needs of poor African Americans. While the need for resources is great and the obstacles difficult, the quality of provision appears to be inadequate. Thus, faith-based institutions dot North Philadelphia but make little dent on poverty and hardship. Religious oratory here, to community development commentator J. Leile (2004), "has moved beyond the walls of the church to the halls, parks, and meeting spaces . . . but with too little positive consequences." In short, the expertise to provide difficult clients with good job counseling, day care, AIDS therapy, and youth support is often not present. As Leile (2004) reports, social service clients in North Philadelphia widely complain about provision of food, shelter, and counseling amid new formal and discursive conventions for identifying need. Of the nine residents we talked to, five said that they must masquerade as "good Christians" to receive food from food pantries and beds at homeless shelters. They note that an informal prioritizing of clients supplants need with deservedness that is measured by religious conviction. At the same time, they complain of poor quality and inappropriate counselors. These workers, four residents note, too often bypass the complexities of client problems and lives as they push religion as general, cure-all principles to social and economic dilemmas.

Philadelphia's faith-based leaders, perhaps unsurprisingly, are anything but saints. Most notably, Mayor Street's appointed head of the city's faith-based initiatives, Reverend Randall E. McCaskill, was recently indicted by a grand jury on charges of theft but continued to draw his $96,000 salary for nearly a year (*American Atheist* 2001). McCaskill, no stranger to politics, helped draw votes to former Mayor Wilson Goode (now a Baptist Minister) and Street in close mayoral elections. His colorful oratory across Philadelphia stated unequivocally who was right for the Mayoralty. Upon becoming mayor, Goode quickly appointed McCaskill to a city post from which he was continuously upgraded in both administrations. McCaskill, with this alliance, oversaw and managed this faith-based initiative for more than three years.

Indianapolis's clergyocracy is also a tightly bound group of institutions that provide flawed services to the poorest African Americans. Even more than in Philadelphia, they affect public policy through dense interconnections with real-estate and educational entities. In Indy, boundaries between religion and public policy have collapsed. This clergy partners with Indy corporate dynamo Eli Lilly and developers to dictate city growth strategies. At the core of this, its touted and influential publication "Religious Institutions as Partners in Community Based Development" was supported by a multimillion dollar grant from the Lily Endowment. This clergyocracy-Lilly bloc recently provided $6.3 million to local university IUPUI to subsidize research that extols faith-based efforts. In 2003, The Endowment also allocated more than $10 million to subsidize expansion of local church facilities, research on the healing properties of religion, and research to make local faith-based initiatives more efficient (see Lilly Endowment 2003).

Yet faith-based resource delivery has hurt many in Indy's impoverished Eastside. Amid the presence of faith-based organizations, poverty and hopelessness worsen. Ten of the fifteen resource recipients we talked to, as in Philadelphia, speak about the need to feign religious adherence to acquire food and shelter. One man, needing food for his wife and two sons, said "everyone professes a faith to Jesus Christ – it's what they want to hear and it allows you to get the pick of the soup and bread." "It's really a sham," he said, "but you do what you gotta do to survive." Many also described to us the job-counseling as pathetic and demeaning. One welfare recipient, in strong tone, declared the exercise a waste of time and degrading. "What they tell me, I already know . . . jobs are mainly at the fast-food places around town . . . so what else is new? . . . but I feel like they're punishing me, the way I'm talked to . . . you think I don't know when I'm talked down to and told my attitude and morals are bad? . . . I hate it."

Cleveland's new faith-based reality is summed up in the workings of City Mission, a powerhouse social service provider for the poor in the tough East Side neighborhood (see Salon 1999). With goals that mirror the new-look social service sector, Christian evangelism permeates all their programs. "We believe God wants us to help the whole person – it's not just soup, soap and salvation" said the Reverend Robert Sandham, City Mission's assistant director (in Salon 1999). In this setting, participants must secure food, obtain counseling services, and obtain shelter by following religious-infused tasks: attending daily worships, Bible studies, one-on-one religious counseling, and peer spiritual discussions. If they are unwilling to participate, assistance is not provided. "Really, the church is the goal," said Sandham. "We want them to have a relationship with God, so they're not dependent on us, but dependent on God, as all of us are. We're real up front with people about that. No one is forced to come here and no one is forced to enter the programs." And if help is provided (e.g. counseling, day care provision), it too is profoundly tinged by Christian beliefs and morals.

Faith-based service provision in these cities ultimately energizes a local clergyocracy and proves a powerful disciplining instrument that, ironically enough, denies its authoritative character. This rule system, presented as a charitable and liberating endeavor, barrages recipients with religion to shape their routine thoughts and actions. But many clients feel the force of authority, and speak openly about feelings of coercion and pejorative labeling. As one recipient of assistance in Philadelphia put it to us,

> we're kind of put on a track of god . . . you know, channeled, cajoled, pushed, whatever . . . and it's all very oriented to finding god and being his loyal subject . . . It's hurting and kind of finger-pointing. It's as though we're the poor and the god-less class . . . you know, the people who don't just need shelter, but saving . . . because of what we are.

To many clients, the reality is clear: these social service providers are more interested in ascribing a character to them rather than alleviating their hunger, homelessness, or need for social services.

CHANGED WORKFARE

We now know that Workfare in 1996 helped create glocal black ghettos in Rust Belt America. Workfare, animated by the articulation of new global times, pushed poor black men and women across rust belt cities into punishingly low-wage and insecure jobs that further institutionalized poverty and marginalization. But Workfare, as now chronicled, has deepened in the Bush years with major consequences as a neoliberal instrument. Conceived as an idea in the early and mid 1980s by conservative policy leaders, it has evolved in perhaps unanticipated ways as a get-tough policy on the "unproductive." In the discussion that follows, to understand this latest programmatic phase, I suggest the reality of three coalescing stages since the policy's formulation: conceptual workfare (1980–85), experimentation-consolidation (1986–2001), and intensification (2001–present). The first two stages, key periods, provided the crucial ground that now allows Bush to intensify and broaden it.

Workfare, perhaps unsurprisingly, is not a new idea (Peck 2001). The notion of work-for-welfare has existed in numerous forms in history. In sixteenth century England, workhouses under the Elizabethan Poor Laws were created across the country for the destitute. Those who could physically work had to earn their right to a "handout." The English Poor Law Report of 1834, codifying this in British thought, stated that incentives of the poor to work were removed with generous, no-expectation relief provided (Peck 2001). Through the nineteenth century, as a result, poor families in the U.K. were frequently separated and forced to work for basic sustenance: shelter, food, and small allowances. More recently, during the 1930s Great Depression in the U.S., a kind of workfare mushroomed in Canada and the U.S. Federally funded camps sprung up that provided shelter and small subsidies (typically 20 to 25 cents per day) to have the poor perform manual labor.

But what is at the foundation of this particular institutional fix? Certain logics appear to propel it. First, as Peck (2001) chronicles, Workfare (like other forms of state sanctioned relief) plays a vital and crucial – but contested – role in regulating labor markets, particularly the bottom rung. Similar to its predecessor, the welfare state, it controls the self-destructive tendency in labor markets for wages and material existences to fall below subsistence levels. This "material floor," placed underneath low wage workers, eases their regulation and assimilation in local labor markets. But what would once have been called a mechanism to aid social reproduction can now be termed something else: a force to assist the making of impoverished poverty spaces. For Workfare now also helps construct isolated and controlled spaces of stigma and poverty as this initiative meshes with neoliberal growth desires and aspirations. This initiative, in this sense, produces not only sufficient bodies of workers to be mobilized, but also the spaces and social relations that they are to strategically occupy as neoliberal growth strategies are pursued.

But at least one other key logic, equally important to ailing local, regional, and national economies, can be said to guide the Workfare initiative. Workfare, now, mobilizes "the unskilled horde" to work in the pivotal sore spot of neoliberalized economies, the dead-end, low-wage sector. This fastest-growing sector of urban, regional, and national economies expands like wildfire across downtowns and retail strips to supplant decent-paying industrial jobs. What is the importance of this "staffing?" Where would current routinized life be, one can ask, without the occupational slots that this low-wage sector now infiltrates (i.e. the circuitry of occupations that include nurses' aides, home health aides, security guards, child care workers, educational assistants, teachers, maids, porters, call-center workers, bank tellers, data-entry keyers, cooks, cashiers, pharmacy workers, poultry and meat processors, agricultural workers, etc.)?

Now to a consideration of Workfare's recent history. Its first phase, conceptual Workfare, involved a handful of influential conservative columnists drawing on ascendant neoliberal thought in the Reagan era to advance the notion of dramatic welfare reform. Reagan's rhetorical attack on the welfare state, in colorful and populist oratory, was crucial. This provided the ground for prominent columnists – Thomas Sowell, Walter Williams, Cal Thomas, Chris Leo, Midge Dector, Mona Charen, and William Safire – to relentlessly pound away at the Great Society's centerpiece: welfare. These columnists, in bold rhetoric, talked about a vast pool of under-tapped productivity from people shackled by a debilitating welfare. Perhaps most vocal, Thomas Sowell offered searing rhetoric about Welfare Queens and welfare-hustling men. At the same time, he started an organization in 1981 – the Black Alternatives Association – to counter the supposed destructiveness of one group, the NAACP. To support his operation, he received pledges exceeding $1 million from conservative foundations and corporations (Aziz 2001).

In 1981, Reagan took the first steps, declaring war on welfare by skillfully pushing and passing the Omnibus Budget Reconciliation Act (OBRA). To eradicate "the ills of welfare," OBRA dramatically restricted eligibility for AFDC, cut welfare payments, and offered inducements for states to develop welfare-to-work programs (Peck 2001). Despite being introduced in the midst of a recession, OBRA removed close to half a million families from welfare (Levitan 1985). But 1980s conservative pundits wanted more. The new centerpiece of their "revolution," the idea of a work-for-welfare scheme, seemingly could not be contained, resonating with writers, media voices, and the American mainstream. T.V. and radio talk shows mainstreamed this notion (e.g. Face The Nation, the McLaughlin Report), influencing an American public suffering from an early 1980s recession, a prolonged ten year economic malaise, and growing anti-immigrant sentiment. The declared enemy was a supposed poverty bureaucracy that taxed public resources, built towering and remote bureaucracies, and left poor people dependent and damaged. By the late 1980s and early 1990s, Congress people, Senators, and Governors were seriously debating what was once a pipe dream: the efficacy of a national workfare initiative.

The late 1980s to the beginning of George Bush II's presidency (2000) ushered in Workfare's second phase, experimentation-consolidation. A first part, a kind of experimental shock treatment, involved Workfare being tried in six states (California, New York, Ohio, Pennsylvania, Georgia, Florida) between 1986 and 1995. Despite much positive media hype, results were soon unambiguous: in the first state to implement this, California in 1986, poverty was little affected (c.f. Udesky 1987; Long 1989). Those that were poor tended to stay poor. At the same time, there was another problem: all six programs inadequately provided resources and services given the use of minimally state-regulated volunteers. Thus, many clients complained of religious zealotry, bad guidance, and poorly trained "professionals."

But undeterred, conservative forces pushed on leading to a second part over the next five years: national implementation. In 1996, Workfare moved from a trial to a national policy: President Clinton boldly proclaimed "the end of welfare as we know it." The centerpiece was time-limited eligibility (maximum of five years on welfare), required work for assistance, and tough sanctions against non-compliance (total cutoff of benefits). Its impacts were immediate and fundamental. Within two years, the nation's number of families on welfare had fallen from 5.1 to 2.5 percent of the national population (U.S. Department of Health and Human Services 2000). Dramatic case-roll reductions were reported in every state in America. Yet its evolution to a refined piece of machinery at the state and local levels still required a steady tinkering with its rules (e.g. appropriate kind of person to provide oversight, participant eligibility requirements, keying Workfare to its rhetorical partner, faith-based initiatives).

By 2000, Workfare was being taken to its next stage, intensification, where it stands today. Bush stepped up Workfare with program specifics now worked out across states and cities, the Fox-popular media fully behind it, and the program nestled in the domain of common sense. At the core of this was re-authorization of the Personal Responsibility and Work Opportunity Act of 1996 in 2002. This legislation led with a new key provision: states no longer had to require Workfare workers to receive minimum wages (see Keller 2002). Bush argued vehemently and successfully that the regulation purportedly punished businesses, particularly "the real heart and soul of American enterprise," small entrepreneurs. Bush pushed this initiative across America, addressing it before rotary clubs, prayer breakfasts, church groups, and the media. Struggling businesses could not pay such wages, Bush repetitiously noted, many would be forced to close and jobs would be lost. Despite the fact that many workfare clients toiled either in public works jobs or national and multinational fast food outlets (see Ehrenreich 2002; deMause 2002), the rhetoric persisted and proved effective.

Bush's deepening of Workfare did not stop here. His 2002 "70–40" reformist proposal, modified by the Senate and made law in 2003, consisted of two new provisions (officially the Work, Opportunity, and Responsibility For Kids Act). First, all states by 2007 were to be penalized fiscally unless 70 percent or more of families receiving welfare were working (Bush also proposed as part of this that the work hours per family required be extended from 30 to 40 – that was not passed). The federal government, paraphrasing Bush (2002), was intent on making states responsible, forceful agents of positive social change. Second, after a participant had been in the program three months, at least 24 hours of work per week was to be in "direct experience." Job training, in other words, had to be dramatically curtailed after three months. Work was to be "real work," Bush communicated, and training and other fluff was to be purged. With minimal opposition from a compliant Congress, this two-pronged proposal easily passed.

In a final new provision, anyone convicted of a felony drug offense was forever barred from obtaining welfare (a last minute amendment tacked on by Senator Phil Gramm) (see Bleifuss 2002). While states may opt out, currently twenty-one states use this provision (see Levi and Appel 2003). As blatant punishment against drug users, ranging from recidivist pot smokers to heroin addicts, government declared that it will aid only those that have demonstrated good citizenship. Defective citizenship and morality, Bush communicated, would not be rewarded and supported by government largesse. The implication: children and youth were acceptable casualties in this war on supposed laziness and bad morals. Innocent kids, caught up in the disciplining of parents for seemingly inexcusable deeds, could be thrust into material deprivation via the punishing of a parent's "lifestyle" and "chosen life course."

Workfare Embedding in Rust Belt Cities

Workfare is now deeply embedded within local frames of knowledge, institutional configurations, and resource bases in rust belt cities. In New York City, the country's largest Workfare undertaking, Mayors Giuliani and Bloomberg have extolled Workfare as the keystone of their mayoralties and the ideal initiative to rouse people from a culture of dependency. "For all able bodied," to Giuliani, "income [s]hould never be doled out without being temporary and without being fundamentally attached to work" (see Butterfield 1998). Giuliani and Pataki also lead the charge to normalize this initiative, discussing it in terms that writer Heather MacDonald describes as "industrial-strength reform." Stereotypes and iconography, wielded like cudgels, offer Workfare as a blockbusting liberatory vehicle for both recipients and the city. This group, as a projected class of unproductive and problematic beings, is to unceremoniously acquiesce to this regime by virtue of their supposed civic unworthiness.

In this context, more than 40,000 people are now employed by city authorities in Workfare assignments (Williams 2002). They clean toilets, tend to grounds, pick up trash, flip burgers in restaurants, and chase off homeless people. The average cost for those workers is $5.80 per hour and $3,600 per year (Williams 2002). An average entry-level unionized worker in New York, in comparison, is paid $18–$22,000 per year. Workfare assignments now also take place in the city's burgeoning fast-food sector: McDonalds, Burger King, Hardees, Wendys, and local supermarkets. These workers mainly cook, clean floors and bathrooms, do counter-work, and scrub machinery. At the same time, an estimated 40 percent of workers hold down multiple jobs to scrape by (see Waller 2002). Working a 50 to 60 hour week is common: twenty hours at McDonalds in the morning and early afternoon can be augmented by working a Hardees afternoon and evening shift the same day.

Thus Workfare's central premise is that participation in any wage labor is a prerequisite for citizenship. A targeted non-working population, whose citizenship is questioned, should acquiesce to this regime for the good of themselves and society. In their predicament, they are to regard any job as a good one, what one Workfare administrator noted as "a step-up in the reality of a working world . . . all around us . . . where every able bodied person should be obliged to work and contribute." This person, expressing the wide-held administrative belief in the logic of incremental-work gain, notes that "such jobs are learning experiences and a kind of attack . . . on deep, disturbing poverty." Thus, every piece of trash picked up, every potato fryer cleaned, every lawn mowed under Workfare auspices propels people to become sturdy, responsible citizens. Such work purportedly initiates a new cultural-behavioral paradigm for these people, one that offers the opportunity for them to assimilate into a rigorous world of waged work.

All ten of the Workfare workers in New York we talked to remain mired in poverty. In this circumstance, more than half identify that this program encodes poverty in their lives albeit with the appearance of working. Labor, in this context, is seen as extra-harsh and ironic, an exercise in anxiety and frustration. "I can't seem to get out from under this [poverty]," one worker in a city park told us. "No matter what I do, and I'm runnin all over the place, the pay is just way too little. . . . Yeah, it's kinda funny and sad in this way." Another worker, in a fast-food restaurant, said "I'm holding down two jobs and getting [welfare] benefits – $80.00 a month . . . they all barely keep us [the family] going. We scrape for food and to pay rent, meanwhile I'm going uptown and downtown just to hold the jobs together." These workers express a desire to work, and to be "out there" being productive. Yet, there was, to our surprise, little hostility and remorse in being in the Workfare program: a resignation to this reality seems pervasive.

Case-workers of Workfare in New York appear to strongly support the program. Three of them, in discussions, acknowledge the reality of clients doing hard, menial work and often staying in poverty, but refuse to indict the program. The complex ideology of cultural-moral conditioning that, as noted, many administrators embrace, is also extolled and transmitted by case-workers. To them, then, Workfare is about fostering both work and the ideology of work. As their comments reveal, they see the program as much an exercise in social and moral engineering as anything else. Workfare, to them, is a vehicle to culturally condition that transcends a simple poverty amelioration initiative. It supposedly builds up and reinforces a set of social and moral codes that is purported to be at the foundation of all stable, contributory workers. As one case worker told us, "Workfare tackles the root of the poverty problem, the character and values of struggling people . . . it's a lifestyle we're changing, we're cultivating how a working lifestyle can be achieved and maintained."

In Cleveland, one of the country's ten largest Workfare undertakings, Mayors White and Campbell have also embraced Workfare as a civic reform. These two eschew the bold rhetoric of the New York politicians but speak simply and solidly on behalf of Workfare: White's one key piece in nurturing the need for "an entrepreneurial city" (see City Club of Cleveland 2006). These mayors, too, have normalized this initiative in routinized discourse (amid contestation and struggle against it) about the need to reform an aging city. This rhetoric thus discursively folds Workfare into the notion of a new, sensible assault on widely recognized city problems: unemployment, poverty, and its supposed corollaries that plague Cleveland, graffiti, blight, crime, and aggressive public overtures (i.e. panhandling, presence of homeless people). The realm of commonsense civic problems, as discursive pieces, becomes tapped and made the enemy of Workfare to bolster this program's legitimacy.

In Cleveland, there are approximately 5,000 Workfare participants. Most work in menial jobs: flipping burgers, cleaning streets, mowing lawns, or picking up refuse. Pay usually ranges from $5.00 to $5.80 an hour with weekly hours set between 25 and 32 (Cleveland Social Services 2004). The average salary, $4,250, falls more than two times below the poverty line for a family of four (Cleveland Social Services 2004). With welfare benefits added in (an average of $65 per week), households still fall below the poverty level. As in New York, many Workfare participants survive by holding down multiple jobs. They, far from being lazy and lethargic, frequently rely on buses and trains in innovative, proactive planning to get from one job to another. When these modes of transport are shuttered or temporarily halted, as discussions with four Workfare participants revealed, taxis and hitching are relied on.

Cleveland's Workfare initiative, also, pushes a cultural-social engineering. "Unproductive" people, blatantly labeled, are put in overwhelmingly menial jobs that are proclaimed to be important experiential interventions into their lives. While much is the same compared to the New York City initiative, there are a number of differences. Most notably, Cleveland's undertaking involves a more explicit "symbolic crusade" that seeks to inculcate participants with the right demeanor and decorum to be fruitful workers. This emphasizes the realm of appearance as much as the domain of skills and abilities, placing importance in effecting a "cosmetic entrepreneurial makeover" that would make these clients more attractive to potential employers. Thus, workers on a daily basis are taught the "right" behavioral codes – deference, compliance, persistence – and superficial markers – dress, manners, style of personal engagement – as an entrepreneurial veneer are cultivated. As one administrator told us, "the appearance and attitude [of workers] is so important to success . . . on this job and futures jobs . . . it'll serve them well . . ."

Workfare participants we talked to here echo the sentiments of the group in New York City. The one difference is a noticeably lessened enthusiasm for participation in the program, which is seen as punishing and dehumanizing. "This work for welfare has me stretched really thin," one worker said. "I've got to drop off my baby at a friend's, catch the bus to work that's a 40 minute ride, make sure I get there on time, and it's all for work that doesn't pay well at all." Another worker spoke at length of difficult work conditions and an onerous boss that is troublesome: "I don't really like the job that much but to survive I've gotta do it. This man [the boss] disrespects me, treats me like I'm nothing, has me cleaning the equipment, floors, the [food] pantry." "If I'm gonna keep doin this, I've got to just ignore it . . . or find work somewhere else." This woman acknowledges her limited job possibilities, but notes that she will leave her current job if work conditions do not improve.

But those currently involved in Cleveland's Workfare are only part of the story. There is also the reality of those trained and weaned off the program, the ultimate goal of Workfare. This group in Cleveland, comprehensively studied by the Center on Urban Poverty and Social Change (2001), appears largely to stay stuck in poverty. The bulk of these workers, leaving welfare entirely as supposed responsible and equipped people, simply lack the jobs skills to secure decent wages. Again, underemployment rather than unemployment is the problem, too few move out of the low-wage, dead-end service sector. At the same time, the Center on Urban Poverty reports, stress on workers continues to be high. For most, health care benefits remain out of reach, child care seriously cuts into earnings, and transportation costs to work are frequently severe. Moving people off welfare, the center concludes, is not the same as moving people out of poverty.

Case-workers and administrators in Cleveland are more critical of Workfare than their counterparts in New York City. On the one hand, their tone about the program's ideals are upbeat and positive. Again, Workfare is about providing both jobs and the cultural foundation to hold jobs. As one case-worker said, "our mission is important, the poor have to be moved to a world of work and responsibility . . . Workfare is all about the rights of people to work and live productive lives." But they also openly speak about the difficulties of implementing Workfare. At issue, they say, is a job market that does not pay good wages to low-skilled workers with people not easily moved from a "welfare culture" to a "work culture." To one administrator, "there are problems with rapid entry of these workers into a local job market . . . at prevailing wages and at current conditions . . . the supply and demand of the thing is troublesome, it's far from a perfect reality. Our goal is to make this transition as smooth as possible."

The New York City and Cleveland cases demonstrate how profoundly Workfare afflicts wage earners and families in these black ghettos. It pushes "clients" into low-wage work in open labor markets with minimum skills provided, too often minimum respect for their circumstances, and a minimum of support. Workers are told to enthusiastically shed "welfare lifestyles" in economically and symbolically impoverished jobs or suffer the consequences of a punitive state and society. Ultimately, poverty is not only sustained, it is institutionalized within the ideological cover of people now working and off the welfare rolls. Yet, poverty under Workfare becomes something much more socially accepted. Now, a materially-suffering people at least work and are conscious of their lack of productivity. The appearance of trying, mixed with the symbolically potent expression of them forced to respond to the supposed brutal truth of their pathologies, carries the day.

Workfare is thus fundamentally paradoxical: it reproduces poverty but punishes these people for this very condition. The reality the program purports to tackle, it seamlessly sustains. But the Workfare initiative continues on as a duality of order and disorder are inscribed in and produced through it. Regimenting the black poor imposes a "work-orienting" order while they are marked as disordered and the dilemma of poverty and deprivation is sustained. Thus, to impose this order, the disorder is posited. And, in the fundamental irony, the dilemma of the disorder, poverty and poor job skills, is never remediated. The program, widely supported across rust belt cities now transfixed by the obsession of new global times and the need to kick "the unproductive" either into shape, or into the shadows, deftly plays to people's fears and anger. It is nothing new that the black poor fail to exercise much control over the state, the machinery of production, and the other established institutions of social authority. What is new is the degree to which this government program and others flagrantly contribute to producing an agreed-upon affliction: persistent poverty.

NO CHILD LEFT BEHIND

Black ghettos in these rust belt cities are also now seared in another Bush initiative, his 2002 "No Child Left Behind" (NCLB) Act. To some an innocuous education reform measure, it has become the new unhidden hand in educational settings that unleashes a "get-tough-entrepreneneurial" wrath on black, poor kids. To its supporters, NCLB is advertised across rust belt states as "an . . . exciting blend of new requirements, new incentives and new resources [that] challenge states, schools, and districts . . . [to provide] stronger public schools and a better-prepared teacher workforce" (Illinois

Board of Education 2003). Bush, proclaiming broken and deteriorating public schools across Urban America and beyond that are unacceptable in new global times, points to the problems of entrenched educational bureaucracies and poor teacher skills and motivations. To fund his notion of an "educational miracle," the national program was provided $1 billion a year for five years in 2002.

At the core of this, Bush casts public schools as blatantly functional instruments in new global times. To Bush, the era of excessive, superfluous education that fails to nurture the essential combo of proper character and crucial job skills will no longer be tolerated, both kids and educators have to know this. So spelled out, this directive works through routinized principles and practices to place "under-performers" (i.e. poor test-takers, poor teachers, poor administrators) in the domain of disreputable. The goal, to Bush, is simple: "a character education . . . children must learn how to make a living . . . [and] intelligence is not enough . . . intelligence plus character – that is the true goal of education" (in Henderson 1999). If public schools fail, he declares, there is an alternative (private schools) that will accomplish this. Public schools, of course, have always been seen as venues to socialize kids into acceptable society. But under Bush, the new world order – globalization – now compels this institution to even more urgently spin out "productive citizens."

NCLB's roots too lie in a curious history – in Bush's dubious intervention into Texas public schools in the 1990s. When governor, his so-called "Texas Miracle" upgraded public schools by instituting a new idea: keying aid and administrative salaries to school performance. This initiative blatantly marketized public schools to spur desired results, using the moniker of "the marketplace of performance" to determine levels of compensation for workers (teachers and principals) and schools (aid). Thus, District superintendents were financially rewarded or penalized (they could lose up to $20,000 in performance pay) based on school results. School principals could lose up to $5,000. At the same time, general funds for schools, this institution's lifeblood (i.e. it pays for computers, teacher aids, music programs, tutors, physical education facilities, and extra-curricular activities) were tied to performance. Standardized tests, not surprisingly, were made the holy grail to measure school performance.

Across Houston, one key test site, astounding reductions in high-school dropout rates were reported *circa* 2000. Teachers and principals supposedly found something that had previously eluded them – incentives to work hard – that enhanced their performance. These workers, once supposedly stumbling through daily work routines in staid bureaucracies, now purportedly had clear incentives to work harder and more effectively. The results were most astounding in the least likely place, Houston's minority impoverished areas. In particular, Houston's heavily poor Sharpstown High School, with over 2,000 students, purportedly had no one drop out after dropout rates of over 60 percent the two previous years. In 2000, this initiative was brought to Washington and aggressively promoted by Bush's new education czar, ex-Houston Education Head Rod Paige.

But all was not so simple: a more harsh portrait is revealed with the situation closely scrutinized. In particular, administrative sleight-of-hand appears to have been at the center of Sharpstown's (and the other Houston schools') seemingly remarkable turnaround. First, many low reading ninth graders in the schools were held back from taking the all-important tenth grade test. This was done under the guise of offering new rigorous "advancement procedures" that kept lesser students from advancing. Second, the category dropout was narrowly defined to include only those who stopped going to school and notified educators. Dropouts, in this notion, had to so label themselves (i.e. as failures) and institutionalize their choice. Third, hundreds of students were incorrectly but conveniently listed as transfers. These kids stopped attending classes (i.e. "dropped out") but became identified as attending other schools. Houston officials now estimate that when these loopholes are closed, the dropout rate at Sharpstown was over 70 percent, an increase from previous years. The Texas Educational Association, after auditing records and finding these discrepancies in 16 Houston schools, was outraged. They recommended that the district rating be immediately changed from academically acceptable to unacceptable.

It is difficult to definitively decipher the motivations behind this fabrication. Venturing an opinion, Bush wanted to make it work convinced of the initiative's soundness and simply did not closely question the findings. Its market orientation (i.e. treating public schools like businesses in need of more efficiency and economic incentives), clearly appealed to his neoliberal sensibilities. Paige's motivations may have been more sinister. Knowing that Bush wanted the Texas experiment to work, and sharing the same values, he was going to make it work, by hook or crook. Or, tying his career trajectory to Bush's, the experiment's success was seen as his ticket to upward mobility. For Paige to climb higher, Bush's experiment had to be shown as successful.

Upon coming to Washington, Paige has changed little in this Texas model. Testing, monitoring, and punishment are at the core of the new national initiative to upgrade public schools: all states must comply. Yearly, states calculate "adequate yearly progress" (AYP) of all schools and school districts to determine their performance based on established annual targets. School aid is tied to acceptable school and district performance. Schools must show improved reading and math scores, test-taking rates, attendance rates in elementary and middle schools, and graduation rates in high schools (in 2004 the government expanded the program and began experimenting with a standardized literacy and math test to all children in the Head Start program. The average age here is four). For schools that fail to achieve AYP in consecutive years, tutoring programs must be initiated and students and their parents must be offered opportunities to choose alternative schools. Acceptable alternatives are public or private schools, with transport costs to both being footed by the penalized public schools.

NCLB Embedding in Rust Belt Cities

Black ghettos in rust belt cities, much evidence suggests, get hammered by NCLB. Most dramatically, NCLB sets up a system of incentives where schools are best served by removing or expelling bad test-takers. If you can't falsely code and categorize the books (i.e. "Rod Paige your school") this is the logical alternative. With school aid tied to performance, poor test-takers can damage school standing and the ability to secure resources and are best purged in the short- or long-term. Short-term removal can conveniently occur over key testing days; long-term expulsions can be meted out over longer periods. The rhetoric of zero tolerance, one central battle-cry of post-1990 conservative Republicans, can be the convenient cover to do this. NCLB, as a class-racial policy formation, ultimately embeds in local educational circuits this structure of identifying and punishing poor black youth.

This "select purging" now widely occurs through disciplinary measures for identified "behavioral problems." Excessive fighting or tardiness lead to temporary suspension from schools, and more serious transgressions like carrying drugs or alcohol lead to permanent expulsions (i.e. the application of stepped-up "zero tolerance"). Principals and Vice-Principals, the key decision makers, operate in a new organizational framework that makes such actions perversely logical. They are compelled to protect their school's lifeblood, money, that cannot go lower without often a serious loss of quality. In response to this, they ironically turn against the most vulnerable students to protect the possibility of providing a more enriching experience for the generic category "student." Invoking zero tolerance becomes the cover, or the mechanism of conscience, to perform the school-protecting deed. But what we see, to McKenzie (2003), is something afflicting: waves of "school push-outs and dropouts [that] pour out of classrooms before their times . . ."

The evidence of this is all over the rust belt. In Massachusetts and its cities (notably Boston and Springfield), suspension and expulsion rates in 2003 soared to their highest point in decades, with African Americans and Hispanics most frequently disciplined (O'Hanian 2003). In 2003, 1,890 students across Massachusetts were suspended for 10 days or more, a 9.9 percent increase from 2002. The reality is the same in Chicago, Indianapolis, and St. Louis. In each of these cities, suspension and expulsion rates in 2003 set all-time highs, with "get-tough" administrators using zero-tolerance rhetoric to push kids out in droves. Nearly 130,000, or 6.45 percent of all Illinois pupils, were suspended in 2002 (Tompkins 2004). In some controversial examples, kids have been suspended for having a bullet in their lunch box, improper use of language, being late to class, and tossing food in the cafeteria. Paraphrasing O'Hanian, if schools want certain kids out, they will find a way.

Prospects for these purged kids, of course, cannot be removed from the dismal economy that traps them in dead-end jobs upon leaving school. Since 2000, America's plummet into the dot.com recession has resulted in 2.8 million factory jobs lost. In the last four years, with sustained deindustrialization,

6 percent of the nation's manufacturing base has closed (McKenzie 2003). November, 2003 marked the thirty-ninth straight month of job losses in manufacturing: the nation has lost 2.9 million private-sector jobs since Bush took office in 2001 (McKenzie 2003). As Bush's post-Keynesian neoliberal economy stumbles along, and his sole jobs initiative has been to cut taxes for the wealthy, these kids bear the brunt of this. While the low-wage service economy – fast-food restaurants, convenience stores, and Wal-Mart like outlets – prospers, more than 85 percent of these jobs fail to provide any health care benefits, child care provision, or pensions (Shulman 2003).

Anecdotal evidence suggests that these purged-from-school kids now help to bolster the fast-food and low-wage service economy in rust belt cities. In cities like Chicago, St. Louis, Detroit, and Cleveland, one readily sees men and women trudging off or clustering at bus-stops in mid-day to access typically poor-paying, dead-end jobs. These economies now rely on racialized adolescents and young men to staff stores. Fast-food mega-corporations like Burger King, McDonalds, and KFC in these locations now have elaborate hire and train programs that rely on these people (see Ehrenreich 2002). Training of these workers to perform the day's menial tasks – flipping burgers, cleaning equipment, mopping floors – is done at many store sites depending upon openings and needs. Now, this population feeds the new downtown economic tiger, the post-Keynesian service economy, as a reliable source of cheap labor for local business. In this context, growing numbers of planners and politicians struggling to entrepreneurialize their cities view this population in coolly pragmatic terms: as facilitative of local economic development (see Wilson 2006).

NCLB's effect on Chicago's South Side is increasingly obvious. The area now has clusters of kids on street corners congregating in afternoons with little to do. This is nothing new in the capitalist throwaway economy, except that increased numbers of these kids appear to be recently discarded public school students. In the Chatham neighborhood, for example, many that we talked to have been expelled from the nearby high school via the new disciplining and "stronger academic standards." "Not much to do, not that people really care that much," one boy said to me. "School?" he says. "I got into trouble, my grades aren't good, before I knew it, I was kicked out." He says, "I'm not the only one . . . many of my friends have been booted . . . [Now] we don't have much to do, pickin' up some work and lots of killin time. The jobs are s___ . . . pay is bad, the things I have to do are just stupid, but it's a little bit of money, so I hang in."

Indy's impoverished Eastside is much the same. The area, staked out decades ago as Indy's most cancerous area by programs, policies, and rhetorical formations, has always stored a multitude of thrown-away young people. Numbers of unemployed or underemployed black men have swelled recently due to the rash of kids expelled from nearby schools. Like the older men, to sociologist Tim Maher (2003), the kids struggle to find their place in a punishing low-wage service economy. Many, to Maher, hold down multiple jobs. Job turnover is high and getting to work is frequently a problem. But the hope of better wages and better working conditions rarely materializes. Most afternoons, many of these kids briefly "hang out" in the local parks and streets. Comraderie with those similarly dispossessed, to Maher, make them feel better about their circumstance. "Life goes on," one expelled kid told us, "but without school and . . . [confined to dead-end] bad jobs, things are just tough."

Despite this, NCLB churns ahead as policy. Rust belt cities like Chicago, Indy, Detroit, and St. Louis now either anticipate or have begun closure of numerous public schools (they've failed under NCLB) for replacement by charter schools. The leading edge of this, the Chicago experience, now aggressively privatizes schools under Mayor Daley's Renaissance 2010 Plan (Lipman 2005). It is closing 60 public schools and opening 68 new charter schools to be run by private organizations and staffed by non-union teachers and administrators. The closed schools, concentrating in Chicago's South and West Sides, affect approximately 72,000 kids. As Lipman (2005) reports, masses of students on the South Side have attended four schools in three years as closures and transfers mark their everyday. While physical infrastructure is re-shuffled, public schools are demonized as failed, subjected to deepened disinvestment, and kids', schedules are disrupted.

The unhidden hand of corporate desires, not surprisingly, drives this. Renaissance 2010 is propelled by the Commercial Club of Chicago (CCC), an institution of the city's top corporate, financial, and political elites committed to entrepreneurializing public education. Its July 2003 report, Left Behind, calls for a total overhaul of Chicago public schools along a choice and market model. Within one year, Mayor Daley announced Renaissance 2010 at a Commercial Club event, and the CCC agreed to raise $50 million for the project. An oversight body for Renaissance 2010 was also established, "New Schools for Chicago," composed of select corporate CEOs and Chicago Public School leaders from the Commercial Club. This "shadow cabinet," so termed by the Chicago *Sun-Times* (in Lipman 2005), was anchored by the chairs of Northern Trust Bank, the Tribune Corporation, McDonalds Corporation, and the CEO of Chicago Community Trust (a major corporate foundation). Chicago Public Schools, creating a new leadership position to spur the project, tapped David Vitale, former vice president of Bank One and current CEO of the Chicago Board of Trade.

A final note about NCLB's relation to black ghetto kids and their communities: through this legislation, more of these kids become coaxed into the military and fight and die or get maimed in Afghanistan and Iraq. A bizarre provision buried in this legislation requires that all public high schools provide the Pentagon with the names, addresses, and phone numbers of its juniors and seniors (see Bleifuss 2002). Schools that fail to comply can lose federal funding. This tack helps the U.S. military-industrial complex gather fighting bodies (with a yearly budget of $440.6 billion) to feed the current wars in Afghanistan and Iraq (disproportionally black and poor) (Shah 2005). But we should not distance this tack from the demise of the Keynesian–Fordist economy. As the new flexible production-neoliberal economy limps along, the military-industrial complex is widely seen in Bush policy circles as a key growth engine (Brenner and Theodore 2002). The "vitamins from warfare" – price stability through control of production, guaranteed markets for "multiplier-rich" arms builders – help underpin Bush's rationale for war. In this way, NCLB and the military establishment collide.

With this information, the military now aggressively targets minority poor kids across the rust belt for recruitment (spending $4 billion on recruiting nationwide in 2004). In 2005, the military hired the Los Angeles advertising firm Muse Cordero Chen to craft T.V. and radio ads targeting African-American communities (*Democracy Now* 2005). Similarly, the San Antonio-based firm Cartel Creativo was hired to create Spanish-language ads targeting Latino populations. Recruitment techniques, *Coastal Post Online* (2003) reports, are typically glitzy and deftly "cultured." Appeals to sports, hip-hop, and a "take to the streets" aesthetic provide the populist veneer that contextualize plays to patriotism, fear of terrorism, and new job opportunities. Now, military recruiters often cruise up to schools in fancy cars, talk on cell phones, and blare music from the likes of rappers Jay Z and 50 Cent. Before the dialogue to enlist, there is discussion of careers, opportunities for travel that the military can provide, and street talk (see Martindale 2005).

In this setting, the tentacles of the military now reach deep into these neighborhoods. Programs like Young Marines, Starbase Atlantis, and the Junior Reserve Officers Training Corps (JROTC) currently dot middle schools and high schools across rust belt black ghettos. More than 50,000 students in approximately 3,000 public schools are currently in these programs (Coastal Post Online 2003). In 2003, 2,000 eighth graders in Chicago's poor black Bronzeville applied for the 140 spots in the Chicago Military Academy. In the same year, over 1,000 eighth graders in nearby Chatham applied for the same spots. Typically, about 45 percent of JROTC students enlist. In 2003, the Army spent $2.7 billion to recruit in schools and communities (double the budget of the 1990s), spending approximately $13,000 to get one kid into boot camp (Pablo 2004). Not surprisingly, the Army and Air Force in 2002 reached its quota of 37,283 and 79,500 recruits in record time (although backlash against the war now makes the reaching of these quotas difficult).

The new military presence in these public schools is not surprising. Financially strapped districts, sapped by the post-Keynesian federal government retrenchment and continued deindustrialization, receive funds from the military to establish these. Financially burdened states and localities, hungry for funds, typically welcome the military subsidy. At the same time, joining the military in these neighborhoods is frequently identified by black youth as one way out of poverty: they are barraged by this theme. But as Brooklyn Council Member Charles Barron (in Pablo 2004) notes, perception of opportunity rather than patriotism often compels military service by black poor youth. Barron's reaction to this has been strong: "I don't want to hear about the statistics and the disproportionate numbers . . . the main reason that people of color join the military is to escape poverty". It is the sheer desire to escape impoverished neighborhoods, to Barron, which makes this such a tragedy.

NCLB, in sum, is a new racial-class apparatus in educational systems with major repercussions for poor black neighborhoods. It is hailed as the lynchpin to culturally and morally re-sculpt kids in new global times. But this educational policy operates as a disciplining, iron-hand that covers over the complexities of low test-takers in these settings (e.g. the influence of parent unemployment, inadequate health care, poverty, inadequate housing, and disinvested and stigmatized neighborhoods). In this context, public schools have become a staging-ground to boldly re-engineer people and institutions to marketize social and cultural climates. A key part of this, poor black youth, are to be inculcated with vocational skills, an entrepreneurial mind-set, and responsible work habits, or they are to be forgotten. Writer Greg Palast (2004), along these lines, describes NCLB as an explicit "hunt[ing] down, identif[ication], and target[ing]" of low- and moderate-income students for either global re-molding or peripheralization. This program, to Palast, "identify[ies] the nation's loser-class early on . . . trap[s] them, then train[s] them cheap" for desired low-wage servitude or casual dismissal.

NCLB is ultimately a rhetorical formation that perpetuates a notion of ghetto cultural and moral disorder. The sheer existence of the program is built around a supposed clear reality: of self-defeating and failing schools in the poor black communities. Its rules and regulations, in bold state proclamation, are deemed necessary to combat this reality. In the process, daily reporting about this program routinely serves up swaths of low-income, predominantly black-attended schools put on "watch lists" and declared failed and "deficient performers." Again, black cultural and social deficiency is seen to infect cities and society that propagates and normalizes the urban racial-cultural divide. In the process, the notion of unworthy poor black households and social spaces becomes, in one more way, casually but subtly reinforced.

THE FUTURE FEDERAL ROLE – BLOCK GRANTS

Finally, what is the future of federal involvement in poor black communities? Will the programs discussed in this chapter be continued? As was chronicled, once the federal government relied on a categorical format (e.g. Food Stamps, Urban Renewal, Section 8, CETA) to effect local physical and social change. These programs, with strict federal oversight, confronted poverty, substandard housing, unemployment, and physical blight by explicitly targeting resources for each of these issues. Now Bush proposes a pattern of government involvement via the use of a different tool: block grants. The future, paraphrasing Margy Waller (2003), is at the moment obvious: here a block grant, there a block grant, everywhere a block grant. From food stamps to job training, Head Start to health care, affordable housing to transportation – it's one dominant policy. And this model is to be carried out by the new providers at the local level: armies of volunteers, faith-based healers, and private-and-corporate "armies of compassion."

This block granting push is now being extended to a new terrain: human service programs. Proposed legislation in 2004–5 would provide states with human services block grants that would supplant Head Start, education grants, food stamps, and entitlement funding for Medicaid with general, flexible funds. Enhanced flexibility and anti-federal bureaucratic involvement are the key buzzwords in this push. Bush also proposes the "superwaivers' initiative that would give states authority to streamline any remaining federal programs and waive program rules if deemed "prudent." What would be eliminated are programs that many see as historically strong and resilient. Head Start, a $6.7 billion program, has been providing nutrition, health care, and early education to nearly one million low-income 3 and 4 year olds since 1964. Medicaid, a 1965 program that pays for necessary medical assistance for low income households, makes health care affordable to the neediest. Food stamps have provided food to low-income individuals directly for more than 25 years.

Bush's block grant plan is deft. Under the guise of state control, he can dump new responsibilities on state governments, as federal revenues via the weakened post-Fordist economy lag (just as states are in an unprecedented budget crisis). States already saddled with stagnant federal aid, dwindled tax revenues from the post-2001 recession and economic malaise, and massive tax cuts, will be told to do more (i.e. assume responsibility for six new government initiatives once done by the federal government, with capped funds provided). The dilemma is that federal aid to states, controlled for inflation and excluding the funding of the prison industrial complex, has declined more than 10 percent between 1990 and 2000 (see U.S. Census Bureau 2002). Moreover, the 2001 recession and malaise cost U.S. states more than 40 billion in revenues (National Governors Association, 2002). In this context, 2005 projected budget holes for Illinois, Indiana, Wisconsin, Ohio, and Michigan are $4 billion, $1 billion, $3.2 billion, $3 billion, and $1.2 billion, respectively.

Political capital would flow from this for Bush: it would allow him to elude responsibility for failure to help households and kids out of poverty and ill-health. As Kleiman (2003) notes, dumping dollars on states means that they could also be blamed for social problems (whoever is seen to be running programs, it is reasoned, can be blamed for program failures). States, then, in the Bush schema, have become prominent resource providers but also fall guys for any political fallout. Yet, in conclusion, all is not easy or deterministic as it seems: this ascendant scenario is not without pitfalls and contradictions. As we learn in the upcoming chapters, the stability and duration of this pattern of federal involvement is open-ended and is currently being contested.

III.

The active black ghetto

6 Ghetto responses

INTRODUCTION

There is an important assertion that has so far been neglected: these ghetto populations have not been passive in the face of this afflictive policy and rhetoric. Assuming passivity in these spaces is easy: media reportage of overt activism has been minimal, the engines of neoliberal policy seem powerful and blanketing, and there has been little documentation in this book so far of resistance to this new policy and rhetoric. Yet, amid a deeper squeezing of these spaces and people, we can detect and now discuss a continuance of both flagrant activism and nuanced forms of resistance. The heyday of overt resistance to city and growth machine policy, of course, was the 1960s, when frequently radicalized churches and activists spearheaded efforts to improve housing, block urban renewal and highway construction projects, and enhance community security. The symbolic center, the Black Panthers who formed in 1966, provided medical clinics, free food daily to over 10,000 children, and schooling that advocated black empowerment (Foner 2002). Today, much has changed, but, as stated above, such flagrant activism and more nuanced forms of resistance persist.

This chapter examines the diverse forms of resistance in these neighborhoods today around one issue: the afflictive global trope. While much resistance and organizing has taken place around the manifestations of this process – substandard and inadequate housing, unemployment and underemployment, excessively punitive schools, physical dilapidation – the new global rhetoric itself has also become increasingly contested. But it is rare, Fairclaugh (1992) notes, that rhetorical formations receive more political scrutiny than the outcomes that they produce. Yet, as I document, awareness of this rhetorical formation grows even amid the sustained neoliberal stranglehold in the rust belt. But, I suggest, difficulties in organizing against this persist, and flow from the assertion's tenacity and its deft seizure of the terrain of mainstream normalcy.

There are some important initial questions. What do I mean by resistance? Who is involved in it? And what are its aims? Resistance, borrowing from Don Mitchell (2000), are acts of people that attack established protocol or understandings and oppose hegemonic normalcy to help create new ways to view, be, and act. They range from small, individual acts of transgression (e.g. mode of dress, kind of speech, contemplative walking) to more aggregate, organized offerings of movements (e.g. rent strikes, anti-gentrification sit-ins). All are interventions into the rhythms of social and political life that re-work social fabrics as either innocuous or flagrant counters. Thus, styles of dress impede and re-inscribe common meanings and re-work identities of youth, walking opens up a city's spatial organization to contemplation and subversive imaginings, massive protest confronts the ills of oppression directly. All are important: they respond to and reject encoded meanings in social life in small or big ways that help establish new paths to see, feel, and live.

But there is an important caveat to using this notion of resistance. My goal is not to socially or politically romanticize life in these neighborhoods. Seeing politics in the mundane and ordinary is not the same as naively inventing imaginary political worlds. Through this notion of resistance, I reveal the complex, elusive, and multi-faceted nature of politics in these settings. In the lived world of communities, as Michel De Certeau (1984) notes, seemingly mundane acts of politics are constant, individually liberating, and always threatening to crystallize into full fledged social movements. These mundane acts, on the one hand, set the stage for the ascendancy of formal, visible uprisings like anti-racism and anti-gentrification protests (i.e. these uprisings rely on new, evolving ways to see and code local life), but also, to De Certeau, such innocuous acts can be individually empowering. People, under the nets of detection, free themselves from debilitating ascriptions that provide new spaces for fruitful living and thinking. For these reasons, I view resistance as complex, diverse, and often elusive.

Who is involved in resistance in this ghetto? The broad scope of this notion I use suggests involvement by many. For example, kids routinely resist, confront, and modify hegemonic culture through performative speech and body performance. Church leaders frequently condemn punitive youth measures like No Child Left Behind and police harassment in scathing oratory. At the same time, adults mentally appropriate and manipulate oppressive spatial organizations in routinized walking; tenant groups fight exploitive and racist housing agents; and men and women listen to blues, rap, and jazz music that fashions different realities to ponder. Local life in these ghettos, it follows, buzzes with near-constant refutations of power relations that often elude the label of politics. Diverse people here, in daily lives, seamlessly or overtly incorporate politics into their routinized practices.

The aims of such resistance vary. Michel de Certeau's (1984) distinction between tactics and strategies is important. De Certeau, concerned with micro-resistive practices in local settings, distinguishes between small-scale, partial, and often experimental tactics and ambitious, blatant political strategies. While tactics manipulate power relationships in ritualized social acts, strategies are advertent and organized appeals to supplant meanings and power relations. To De Certeau, tactics are a kind of subterfuge that challenges power as dimly recognized forms of politics. They may or may not be calculated, but always work through accepted social practices (e.g. walking, making and listening to music, talking) to advance political challenge. Strategies, in contrast, involve subjects consciously applying their will and power to mobilize masses of people to drive for change. To gain converts, scripts of good, bad, villain, victim, and salvationist are created and deployed in a politics of trying to persuade and organize.

These rust belt black ghettos thus contain, on the one hand, resistance that is small-scale, modest in ambition, and embedded in taken-for-granted life. This throws off repressive meanings and understandings that can provide individuals with something important: better immediate qualities of life. For example, acts of walking, seemingly innocuous, allow individuals to seize a geographical field organized as a grid of power that can be imaginatively re-made. Such walking, far from being a delusional act of re-representation, is a creative and nourishing re-reading of reality. Similarly, the widespread and seemingly non-political act of "street talk" can "hollow out" oppressive meanings in terms and re-fill them with empowering and self-satisfying meanings. Under cover of seeming innocuousness, it can critique oppressive codes and re-make words to fruitfully allow a speaking of one's truths. Such tactics typically elude the vision of the police, planners, corporations, and the state to exist, in Don Mitchell's (2000) words, "below [the] radar."

But there is also flagrant resistance. The aim typically is to increase the flow of resources into communities and to institutionalize this. Churches, tenant groups, block clubs, and the like seek, in De Certeau's terms, a calculated hit on authority. People and organizations assemble and seek converts to thwart the likes of housing policies, bank and Realtor practices, policing strategies, established ascriptions of African Americans, school regulative structures, procedures at City Hall, and practices of politicians. This resistance, unlike tactics, is aggregate rather than individual and always looks to increase its numbers. It thus assaults the base of established power, opposing rather than ceding significant power to it. Even as power and vision are often stultified through unavoidable use of borrowed language and inherited conditions (i.e. meaning-colonized terms and conditions), established power is attacked.

A key point to understanding this flagrant resistance, is that a frame of history and social setting continuously bounds and defines it. An important manifestation of this: as society's expectations of "a people" set in "a place" change, so too do the aims of resistance. In the 1960s, notions of equality across race and class infiltrated common consciousness, led by the institutionalizing of the Great Society. Resistance in these ghettos became aggressive and frequently targeted local and societal structures that propagated inequality (e.g. raced and classed federal housing policy, the logic of inner city deindustrialization). But this has evaporated with the public's marginalizing of concerns about equity and equality. Now, in the neoliberal era, a staunch neoliberal and revanchist sentiment flagrantly denigrates poor African Americans that humbles some efforts at bold resistance. Today, the drive is frequently more modest: to secure more housing funds, reduce racial profiling by police, and acquire more businesses and jobs, which replace once frequent calls to restructure racist institutional practices (e.g. bank redlining, Realtor steering, exclusionary zoning).

RESISTANCE TO THE GLOCAL GHETTO

Resistance to the global rhetoric is growing across the rust belt. Both tactics and strategies now tussle with this ideology's legitimacy in numerous ways. First, as we see in this section, a range of seemingly innocuous tactics (e.g. walking streets, modes of verbal exchange and communication) subtly challenges this rhetoric's content. This is done at the individual and small group level. Second, there is the overt and flagrant challenge of the rhetoric and/or its outcomes. Such social movements sometimes directly challenge the core of the ideology (i.e. the realness of globalization and the need for cities to push what has been identified across the rust belt as "the business agenda.") but more often fixate on its manifestations (i.e. the inequalities of the restructuring itself). While tactics are small scale but frequently potent, e.g. each contrary footstep undermines the power of those who seek to control the spaces of the city, strategies are organized broadsides which unabashadly delimit their numbers, followers, and bases of resources.

Tactic and Cultural Politics As Resistance

The realm of tactic has been and continues to be important in these spaces. Today, a fertile terrain for this in these ghettos is music. Two dominant idioms, rap and hip-hop, are the leading edge of organizing new ways to feel, see, and act. Much, of course, has been written about the frequently overlooked therapeutic and identity-empowering aspects of this music (see Kelley 1997; Lott 1999). It is seen to foster individual and group consciousness that mixes self-pride, non-sublimation to authority, and the aesthetic of community solidarity. But less recognized is that this music, commonly seen as recreational and innocuous, presents listeners with alternative meanings and ways to see "under the radar." Ice Cube, offering caricature, nevertheless notes this music's importance for African Americans: "rap for white kids ain't nothing but a form of entertainment, but for black kids it's a strategy on how to maneuver through life" (in Lott 1999: 103).

But, more than before, this music now embodies resistive content to the new global rhetoric in these rust belt ghettos. In Indianapolis, Chicago, Cleveland, and St. Louis, rap and hip-hop music is made and consumed that, first, visibly counters this rhetoric's afflictive manifestations – rampant unemployment, deteriorating housing, exacerbated hopelessness, and struggling organizations. Second, it frequently targets key rhetorical themes in the global rhetoric for ridicule and subversion – the obsession to manufacture productive and civic people, the drive to convert schools to strict vocational centers, and the push to infuse downtowns with one version of high culture.

Thus, this music often does double political duty, tussling with both the manifestations and the heart of this global rhetoric under trying and difficult conditions.

In Chicago, self-made rap and hip-hop permeate the streets and public spaces of the South and West Sides. A welter of performative rappers push this micro-politics; one well-known artist, with captive local audiences in clubs and streets, is Tony Green, a.k.a., Paraplegic MC (he has muscular dystrophy). His music, richly textured and lyrical, invokes the hard reality of broken dreams and ambitions of poor black youth for kids to feel and contemplate. For example, his signature song, "Pieces of Dreams," comments on the hardships of his place of birth and upbringing, the Robert Taylor Homes. Paraplegic MC chronicles the pain and suffering induced by joblessness, loss of hope, racism, and socio-spatial isolation. The Taylor Homes are the proclaimed epicenter for a forgotten and neglected community. Neither syrupy nor straightforward, the music evokes these images set against the frame of once-imagined individual ambitions and potential possibilities.

But Paraplegic MC's lyrics run deeper. They also target two central themes in the global trope that now permeate Chicago: the obsession that all people must be productive and civic to feed neoliberal designs, and the drive to gentrify and upscale the city. These lyrics confront the global trope as an oppressive and punishing formation. Black youth, in theme, are "aimless faces to be engineered for responsibility and integrity," black men "programmed . . . to march off to the fast-food economy," black and Hispanic neighborhoods "disposable things . . . that get in the way of building their playgrounds and having fun," and the downtown "nobody's place but the masters of control." In blunt but fleeting images, Paraplegic MC offers a world of corporate power, state hegemony, and a damaged black community. To be sure, there is also commodification in this music, most notably in frequent assertions that "the rap and the beat is the thing" and "the man can take us to harmony." This music, then, like much rap, is a hybrid of things (i.e. of commodification, social commentary, and lyrical convention that is fundamentally oppositional but never completely so).

A host of other rappers – local celebrities and kids – write and practice this music on these Chicago streets. One of them, calling himself "Tupok 2," told us "the [my] view of life, the country, the city, the South Side, it's all in the music . . . It's about how we're in these neighborhoods and forgotten by the people in power . . . there's sex and sexuality . . . in the world . . . in the lyrics . . . no doubt about it, that's the 'hood . . . that's in all of our songs . . . but me and others are lenses onto the scene, truth speakers, we say what others can't say." Another rapper, "Big D," calls himself "the visionary of what's really going on." He says: "I like to rap about families, men and women, the neighborhood, the state of the world". Like Tupok 2, Big D casts himself as an agent of change, someone who calls out for others to notice what is right, wrong, reality, and fiction in his neighborhood, city, and society.

Indy's black ghettos also have kids making and listening to rap and hip-hop music. One local artist, Da Black Don, blends anger and vision that flow from his lyrics like bombshells. Two LPs, "Americaz Most Hated" and "Death Till Dawn," speak about being born and raised on Indy's poor Eastside. A common theme courses through both offerings: the complexities of human survival in a forgotten neighborhood. The song Dreamelation, a startling exposé, speaks of a black man's disillusionment and everyday material struggle set again a background of an ominous, dominant thunderstorm. Similarly, the song "West Side" resonates with an anger about a person's plight of "runnin out of time" in a tough, death-glorified neighborhood. Another song, "The Good Life," is bitter commentary about the illusion of youth being able to avoid gangs and death in this community. Here, he says, "too many [people] lay in the ground . . . and in the cemetery."

Da Black Don's lyrics also do not spare the global trope in ultra-conservative Indianapolis. One key theme is torpedoed: the new disciplining and punishing of black minority youth. "The Good Life" speaks about the isolating and managing of "the black poor contaminant" by invoking a frightened black teen pulled over by the police. "Playin with the Savior" – the police – is the prop to reveal the fear and shame of being affixed a throw-away in relation to "more civilized" people and spaces. Images of being beaten, leading to the "near death of me," offer a view of a brutal and repressive state in the service of real-estate capital. Again, the music is an unmistakable hybrid of commodification, social commentary, and social convention. "The streets," for example, are a place of struggle, exotic hyper-mysteriousness, and mischievousness. Just as the media continuously struggle with the tension to demonize and commodify "the streets," with no end seemingly in sight, Da Black Don too struggles with thematic ambiguity.

The streets of Indy's Eastside are also littered with kids rapping to large groups. Three rappers we talked to were excited to discuss their music. They were quick to mark their music: "gangsta rap," "east coast," "west coast," "hardcore," "underground," "political," or "general." Two of the kids spoke of their music as "a mixed bag of things," the other labeled his as simply "political." But despite these clear distinctions, these kids told us that all of it is designed to be political and speak truths about the world. To "Benny," his music and hip-hop itself is all about "the struggle and anger about our lives and our neighborhoods . . . the need to move out of a strait-jacket . . . and keep hope alive." To "Alex P," "it's an authentic music that alerts the street to what's really happenin . . . we give em the gritty reality of life in the neighborhood and who's behind it."

These communities, through this dialectic of performance and response, are currently a hotbed of alternative social and cultural imagining. Some of this, of course, is politically progressive, other stuff is weird, and still other stuff regressive. Thus, some spoke to us like radical political economists, talking about a city obsessed with promoting gentrification and "rich people's culture" as tools of business interests. Their neighborhoods, they suggested, are throw-away places out of the public and private concern. Other kids were different, hinting at or speaking openly of ethnic and religious conspiracies. Two kids talked to us, for example, about ghettos and the destructive role of koreans and "business-intellectuals". What is going on here, I believe, is that rap and hip-hop, breaking apart established ways of seeing, fill this core with a range of alternatives for common consumption (hence the diversity of political content). They do this by drawing on and melding together other resistive, and widely varying, discourses that circulate across these spaces. To name but a few, Islamic-religious narratives, church oratory, social welfare ideals, and Black Panther rhetoric become drawn on as the new sources for symbolic re-inscription.

Another kind of resistive tactic in these ghettos also re-codes meanings: socially constitutive walking. This common act, widely seen as mundane by many inside and outside these spaces, enables a "socially clarifying" kind of politics. To De Certeau, walking "appropriat[es] . . . the topographical system on the part of the pedestrian . . . [it propels] a spatial acting out" to allow people to configure new "differentiated positions." His "long poem of walking" manipulates spatial positions through something crucial: "calculated action." This act thus confronts hegemonic spatial forms and allows a reconfiguring to trigger alternative imaginings. Power-laden physical forms and social relations, eyed and interrogated, set the stage for transformative thinking, living, and concrete counter-response. These rounds of footsteps, paraphrasing De Certeau, are difficult to control and manage and are seldom innocent.

Numerous people use tactical walking in these rust belt ghettos to resist the global trope. For example, youth gangs employ this to help build two resistive things: (1) a different, more nourishing social structure and (2) a supportive, underground economy whose jobs offer relatively expansive career trajectories. Strategic walking is one task in this multi-constitutive project. First, it helps construct and reinforce a gang social structure that, for all of its controversy, imbues kids with rules and rewards that supplant malaise and hopelessness. Second, it helps provide decent-waged jobs – working in gang-owned stores and shops, selling drugs, peddling goods – to fill the gaping hole left by a disappeared formal economy. These points are not meant to overlook or excuse the frequently tragic participation of gangs in violence or illicit activities. But they show that the growth of gangs reflects a kind of counter-political response and adjustment to growing poverty and disillusionment. As society institutionalizes the destruction of these ghetto spaces, it is not surprising that this counter-political form emerges.

But how does gang walking nourish this social formation? Simply stated, it seizes damaged and harmful spatial grids (already created and coded), helps members re-code and imagine alternative socio-spatial patterns, and facilitates a re-use of these crevices and spaces (streets, blocks, alleys, parks, open spaces). New patterns of understanding and community use follow. In the process, harshly policed streets, people-bereft parks, and dead retail strips become zones for desired activities: youth gathering, regulatory avoidance, and vibrant entrepreneurialism. Debilitated and afflictive spatial forms and activity zones are ultimately replaced by a social milieu that cultivates a sense of adroit, active, entrepreneurial beings. The outcome: negatively scripting and punishing spaces are transformed into positively signifying, nurturing terrains. Kids, in search of buzz, excitement, and decent lives, do so under supportive conditions that they help define rather than living within onerous, imposed realities.

The Black Disciples in Chicago illustrate this. They now provide work for hundreds of kids in legitimate (day-care, laundromats) and illegitimate (selling drugs) endeavors that promise upward mobility for the ambitious and hard-working. As Chandler (2004) reports, they run businesses that are the envy of Chicago's regular and shadow economy: their distribution network brings in as much as $300,000 a day. Its sales territory, divided into clearly defined franchises, involves members paying dues and "taxes" for the right to sell. An annual company picnic reinforces bureaucratic rules and social structure. Its chief competitor, the Gangster Disciples, adopts a similar pyramid-style organization with a CEO-leader on top. A board of directors holds regular meetings and is chief consultant on business operations. Finally, both gangs have a kind of benefits package. The gangs pay annuities to families for those hurt, killed or incarcerated, pay for attorneys and bail, and provide medical costs for members. Members, typically high-school dropouts or expellees from school, are required to spend 20 to 40 hours a week working.

Both gangs also provide a social structure for youth that rejects the pervasive "retreat-into embarrassing poverty" lifestyle. The everyday insecurity that chokes too many kids here gets supplanted by assertions of enterprising, assertive beings. "Victims," as told to us by gang member "Toby," get rescued and become active agents. These gangs, run like a mix of business enterprise and social club, recruit "potentially changeable" and "loyal" youth and place them in a sturdy social formation with the expectation of participation in economic activities. Members work and are elaborately fitted into this social formation. Like most gangs, and contrary to popular belief, there is high turnover: members are not required to make a pledge of membership for life. An estimated one-half to one-third of members stay for one year or less (see National Youth Gang Center, 1998).

In this context, strategic walking is crucial to gang sustenance. These gang members ritually comb streets in a process of surveillance and re-configuring of spatial organization. Mobile youth eye the prevailing hegemonic grid, continuously re-make it as their own, and, paraphrasing De Certeau (1984), turn its seemingly sturdy state into shadows and ambiguities. In this process, dilapidated buildings are functionally re-imagined and used as fortresses, gritty and decayed streets are re-imagined and converted to lively social zones, and marginal storefronts are re-conceptualized and re-used as businesses and entrepreneurial "hot-spots." At the same time, strategic walking allows monitoring of this re-made terrain because it is an ascribed anathema by authoritative sources (the police, planners). Here the re-made terrain and the hegemonic spatial grid are at odds with one another, and potentially clash. Walking, it follows, is central to this surveillance to preserve the ideals and spatial form of the resistive politics.

Another community group, neoliberal-resistive black politicians, also use tactical walking to help thwart the global trope in these ghettos. These agents for community change, many of whom are widely seen as more mainstream than gangs, seek to mobilize collectivities to upgrade community (e.g. secure the likes of more affordable housing, more decent jobs, more public resources, and better access for residents to health care). Tactical walking works at two ends: it is both a self-enlightening measure and a theatricalized, communicative prop to validate this politician's projection of a benevolent identity. The first provides the grist for the politician's raw conceptualizing of politics, the second helps stage and illuminate the veracity of the political position for constituencies and the public to digest. Let me explain.

This walking by these politicians, first, allows a critical "eyeballing" of the community and the world that fires their political imagination. Key neighborhood elements – streets, corners, social organization, and economic structure – become eyed and interrogated for fairness, utility, and value. Ultimately, this deciphering and verbal engagements with adults, youth, and the elderly help to understand community processes. Thus, many progressive politicians in rust belt cities use walking to confirm their suspicions of the myth of dysfunctional subcultures and lethargic people dominating these communities. The proof, found in walks through and decoding of neighborhood space, fuels the resistive political posture. Ultimately, it leads to aggressive narratives about the real problems in these communities (i.e. the likes of ghetto abandonment and policy marginalization by city governments and business communities, which can resonate with residents and others).

Second, walks are used as effective theater to highlight the veracity of the politician's counter visions. They are used, like sound bites, to reduce and freeze a person (the politician) in space and time, to display a concerned individual reading and deciphering clues in troubled communities. A penetrating traversing of space is offered for all to see, part of a strategy of political constructing. This walking thus conveys something crucial: boundary-busting people who defy barriers and obstacles to get at brute truth. Ghetto space, a key part of the set-up, is an offered essence in need of meticulous deciphering. Its complex truths, scattered across this terrain, are shown to be unearthed by mobile, shrewd, interrogating politicians. Such walking counters the idea – fervently asserted by De Certeau – that this resistive practice is always "below radar." Here, walking is a significant practice as a key political maneuver in the politician's arsenal.

Prime example are politicians like Congressman then U.S. Senator Barrack Obama of Chicago, Mayor Sharpe James of Newark, activist Ken Bridges of Philadelphia, and Black Panther leader Mmoja Ajabu of

Indianapolis. While their politics are different, each makes walking and commentary about places central to their political oratory. Most fundamentally, these walks provide the material for the production of claims that run counter to hegemonic assertions. All are notoriously avid and keenly observant walkers' and this "feeds" their most frequent proclamations: Obama narrates the "truth" of city-neglected neighborhoods, James the "reality" of resource-starved inner cities, and Ajabu the blunt "truth" of racist city surveillance and policy brutality. Their walking, a process of reality assemblage and data-gathering, is a kind of maneuver "within the enemy's field of vision" (De Certeau 1984: 37). It seizes capital's and the state's power-laden grids, as re-codable objects, and re-makes them for their political purposes.

At the same time, these walks are frequently publicized and made effective theater to deepen these politicians' benevolent identities. The crucial act of creating vibrant counter-politics, to belle hooks (1993), requires a continuous constituting and re-constituting of credible leader identities in common thought. To Wilson (2005), making such identities by "putting these people in motion" as ceaseless walkers and spatial interrogators is a common ploy. Press releases of these politicians in Chicago, Philadelphia, Newark and Indianapolis, for example, often have them courageously cutting a swath through decayed and decrepit blocks in expeditions to find truth. Baraka is narrated as robustly combing the recesses of Chicago's South Side, James cuts through the thicket of the Fifth Ward's darkness, and Ajabu forcefully strolls through the decrepit landscape of Eastside Indianapolis (see *New York Times*, 1998; Law 2000). Again, countering the notion that resistive walking always avoids political detection, this consciously illuminated act is crucial to the resistive politics at work.

Strategy As Resistance

In rust belt cities, "action" in the realm of formal resistance ("strategy") against the global trope has been pervasive. This organized resistance, with varied success, has galvanized around two themes: the exaggerated contention of a powerful new global reality and the response of cities to this supposed new reality. While the former attacks the truth or influence of a supposed new global world order, the latter accepts the notion of new ominous global times but assails the responses of cities to it. These are very different: One posits globalization as exaggerated and a politically expedient contrivance, a political construct, the other takes aim at the inequities perpetrated via responses to a supposed new reality. It is the second theme, the dilemma of local response to globalization, that dominates this kind of mobilization and resistance. At issue is the reaction of cities to this (i.e. restructuring cities in ways that makes the black and minority poor bear the costs).

In Cleveland, for example, forceful protests have recently arisen over the city's clean-the-downtown-streets policy that harasses and dislocates the city's 3,000 homeless (who are disproportionally black) (see Boland 1999).

The police openly remove these people identified as eyesores and obstacles to livability and upgrade in the downtown. This policy, to Northeast Ohio Coalition For the Homeless Director Brian Davis, reflects the drive of local government and the business community to clean up downtown's image and go global. According to Davis, globalization is the new reality across America and the world, and Cleveland wants to be part of it. In this process, he submits, Cleveland's businesses and corporations, substantially footloose, demand a sanitized downtown to attract needed "human infrastructure" and support investment. To Davis, the black poor bear the costs of this response: they are hurt by the dual unfolding of globalization and a city's perverse reaction to it.

In this case, a successful counter-response has followed using a key strategy: litigation against the Cleveland Police Department. On behalf of four homeless people and the Northeast Ohio Coalition for the Homeless, a lawsuit brought by the ACLU of Ohio went to trial February 18, 1997. The suit alleged that the city was engaged in unconstitutional activities to sanitize the downtown area. The Circuit Court seemed about to reverse a long-standing judicial position in Ohio – that the constitutional rights of communities to enforce safety and nuisance ordinances took precedence over the constitutional rights of the homeless – and this led to an out-of-court settlement. The settlement, reached in 2000, declared that Cleveland police would not harass or arrest homeless people if they were doing nothing illegal. In the settlement, the rights of the homeless – "sleeping, eating, lying, or sitting on public property" – were to be protected (Kropko 2000). Brian Davis called the settlement a victory in the battle to protect Cleveland's poor, and potentially the poor in other cities, from being pushed aside in the ongoing "globalizing" of city downtowns (Kropko 2000).

At the same time, protests across rust belt cities have ensued over the planning tools that help foster city restructuring, particularly tax increment financing and historic preservation. These state tools are said to manufacture patterns of investment and building that are grossly inequitable and devastate black ghettos. Tax increment financing (TIF), a now established staple across numerous rust belt cities, collects local property tax revenues and plows them back into their neighborhoods of initial extraction. Cities, this way, hand-pick neighborhoods that "need" or "deserve" a boost in public-sector investment (e.g. gentrifying neighborhoods, high-tech corridors). Historic preservation, similarly, designates neighborhoods as beneficial to city civic and cultural affairs and provides them with property tax reductions for new investment in housing and a historic-upscale cultural status (Wilson 2004).

In Chicago, the leading city for TIF use in America (over 130 such zones now exist), resistance at key sites is ongoing. In the low-income Pilsen area near downtown, for example, a strategic TIF has helped fuel a pocket of expanding gentrification. Resistance has been fueled by the ability to create and place a rhetorical formation – a save neighborhood discourse – at the center of resident consciousness. It has involved, first, an intense grass-roots organizing campaign within local churches, parks, and informal gathering spots. Residents in these settings are alerted to the reality of dramatic changes in their neighborhood and possible displacement from more affluent in-movers. Second, resistance has involved formal marches on City Hall and at redevelopment sites to impede actions. A nexus of organizations, coalescing into a vitriolic coalition, deploys a simple but effective line: a city-led transformation in Pilsen today could easily spread and engulf much of the community.

In one important sense, this protest has failed. The TIF has remained intact and gentrification has gained a core foothold. But in another sense, the protest has been successful. It has raised the consciousness of both residents and the general public about this issue, and efforts to blunt gentrification beyond here have proven successful. In Pilsen, now, activism against gentrification is common and aggressive. Activists closely monitor the neighborhood, let potential developers know that their projects (and new tenants or homeowners) will not be welcome, and advise them about the risks of proceeding further (i.e. demand for their condos, coops, and apartments could be problematic). This resistive approach, close monitoring of gentrification and communicating the possibility to developers that in-movers could be "shut out" or harassed, is a surprisingly underutilized tool across urban America. In Pilsen, now, this crystallized group consciousness and community intervention alarms developers and mitigates the desire of real-estate capital to gentrify.

But formal mobilization against the global trope has taken another tack: challenging globalization's truth status. The reality of globalization, confronted and ridiculed, is presented to the public as something hideous: a powerful and working-class damaging "politics of ontology." Here globalization is an exaggerated product to assist the designs of prominent businesses, developers, banks, and speculators. The state and local media, anything but innocent, do the business community's bidding. In argument, growth machines, seizing upon a national rhetoric, foist a notion of a new world and the city's need to respond as an expedient invoking to meet economic ambitions. The notion of globalization, it is asserted, gains respectability through a fervent articulation by established pillars of the community: prominent economic and political leaders.

There are many examples in Washington D.C., for instance, homeless activists (Mayday D.C.) buoyed by this stance have recently protested city policy that has hurt the black poor Columbia Heights community. Activists have galvanized public sentiment and other institutions (Washington Inner City Self Help, Latino Workers Rights Organization, National Coalition For the Homeless) to occupy an abandoned building in downtown Washington and march downtown. The leaders, former occupants of a community facility that fed and provided other services to the community's struggling and homeless (Olive Branch House), led the organization. In a highly symbolic act, they issued six demands to the City Council and Mayor's Office rooted in a discourse of a foisted global politics imposed on Washingtonians.

1. That the city commit to a program of at least one-for-one replacement of affordable housing lost to developers.
2. That the city immediately end warehousing of vacant, unused city property and instead place it in a land trust to protect it from predatory developers, real estate speculators, and parasitic "community development corporations."
3. That public officials commit to providing a full range of services including supportive housing for the disabled, treatment on demand for those with drug and alcohol problems, and assistance in locating and securing both housing and employment. An immediate commitment to right to shelter must be reflected in the 2002 Reform Act for Homeless services.
4. That the city pass a living wage law guaranteeing working people a wage that allows them to support themselves and their children.
5. That La Casa remain open and be given a permanent site that will allow it to expand its services to Latina women and families.
6. That the city provide Olive Branch with a leased space that will allow it to restart the services it has had to suspend due to its eviction.

The rhetoric, attractive and incendiary, attacks the sense of globalization's truth status. As Mayday D.C. organizer and former Olive Branch resident, Jamie Loughner, said (2002, 2004):

> we ma[ke] residents see that globalization is a construction . . . it is a tool used used by others . . . to advance their own interests . . . Poverty and gentrification are overwhelming the residents of Washington D.C. . . . Over the past ten years the number of children living in poverty in the district has increased 29 percent . . . Meanwhile the city has done nothing to create and preserve affordable housing as thousands of units are lost to gentrifiers . . . the city is cutting back human services like hospital and shelters . . . taking their resources elsewhere . . . under the name of new global needs.

This particular case was unusually effective. First, the occupation and march received considerable publicity in the local media. T.V. and radio reportage displayed articulate and thoughtful occupiers of a building, determined to help their community. Second, the city responded tentatively but affirmatively. To be sure, there was much talk and inaction: the city delayed any serious action for weeks. But an important concrete step was finally taken: the La Casa shelter was saved. In the process, the community's major drug treatment facility and homeless shelter was preserved. This was, to Jamie Loughner and many residents, a major victory. The result: the burgeoning homeless and drug-addicted population had a place to go and be served. At the same time, in the realm of the symbolic, it meant local residents could make a difference in community redevelopment and the global trope could be resisted.

A similar successful movement occurred in New York City in 2002. In a highly publicized and effective resistance, hundreds of demonstrators rallied to protest emergent gentrification and displacement in Black Harlem. The city-wide rally, initiated by the Harlem Tenants Council to coincide with new rent hikes set by the Rent Guideline Board, built on a growing rhetorical outcry against community gentrification. It featured Harlem tenants, activists, and civic leaders coalescing to form the Harlem Coalition to Fight Gentrification. Protestors marched from two locations, 116th Street and 124th Street and converged on the Adam Clayton Powell Building. The rhetorical core was the notion of an exaggerated hyperbole about new global times that was being used to propel gentrification. The rhetoric, brash and luminous, mobilized thousands of people.

Again, the rhetoric stripped the global notion of its truth status to re-frame understanding of current community development. Activists talked boldly of "greedy politics" and the state's strategic use of the notion globalization. The public, it contended, was saturated with invocations of the new global reality and how gentrification and historic preservation were needed in these times. Promoters of gentrification, it was said, worked constantly to advance their agenda with Harlem residents having to pay the price. To Nellie Bailey (2004):

> Over 60 percent of Harlem residents liv[e] below the federal poverty level as well as moderate-income renters . . . all are angry . . . all are unable to pay $2,000 for a one-bedroom apartment . . . we have to see city policy at the root of this . . . they [the City] want the entire Borough gentrified, the new global thing and all dictating this . . . Harlem residents would be kicked out to live elsewhere.

The protest and march was widely covered in the local media. Perhaps most importantly, it set off a renewed activism in Harlem that has led to subsequent clashes at gentrification sites and the checking of this process. Currently, the Harlem Tenants Council has a new broad-based agenda: it organizes educational forums, provides free legal counseling, builds ties with Harlem churches and businesses, and organizes demonstrations to draw attention to a community housing problem. Since the highly symbolic and visible rally and march in 2002, HTC has stepped up its actions. In the process, other politicians and organizations have become active to check Harlem's gentrification: City Council Member Charles Barron, State Senator David Paterson, the Latino Workers Center, Harlem Fight Back, and Harlem Operation Take Back.

THE OUTCOME

Resistance to the tools, authority, and meanings in the global trope in rust belt cities is ongoing with mixed results. If the present is an accurate indicator of trends, this resistance will continue even as growth machines work to maintain their influence. But it is in the realm of formal resistance, where people explicitly wear an insurgent identity and engage in flagrant acts of politics, that the dilemmas to organizing are particularly severe, and I conclude this chapter by focusing on this.

Perhaps most difficult here, I believe, is that the global trope is anything but a sitting target to be easily labeled and rallied against. Like many deft rhetorical formations, it has proven to be a highly flexible and adjustive construction as points of weakness are identified and corrected. For example, this rust belt rhetoric today in a circuitry of cultural production subtly "shades" to encompass new civic demons and problems to stay contemporary and stave off potential criticism. To illustrate, in Chicago and Indianapolis there have been recent bouts of anger and organization against city policy (i.e. police crackdowns in poor minority neighborhoods, racial profiling), and casting the black poor as the city's most virulent demons has momentarily waned (see Wilson 1996, 2005). "Immigrants," particularly Hispanic and Mexican people, ascend as the new fall guys, which corresponds with their recent influx here. With the heat subsiding, this villainizing of "new immigrants" has not gone away, but the black fall guy, again, re-assumes centerstage. Making enemies that advance political projects, Linda McDowell (1995) reports, is a fragile and nuanced undertaking.

This organizing is complicated also by the nature of the resources available to enact resistance. Paraphrasing Don Mitchell (2000), it is difficult to dismantle the master's house using the master's tools. Mitchell means that the language and understandings that must be negotiated to begin re-imagining and confronting the likes of the global trope tend to be the property of the state and capital. "Stocks of truths" embedded in this language and oratory, invoked by the state and capital, often coalesce into something formidable in places: a thick, dense "state of consciousness" that impedes political mobilizing (e.g., it authenticates the reality of new global times, offers its supporters as truth-sayers, portrays opposition as misguided and futile). The terminology of globalization, at the core of this, contains influential meaning-infused terms: "the global," "the new entrepreneur," "the business community," "the ghetto," "the black underclass." Each term, a dense repository that traps meanings in what Mikhael Bakhtin (1981) calls its archive, seamlessly spills out understandings for common consumption that support this offering of globalization.

In this process, then, the global trope often firmly seizes the terrain of objectivity and neutrality. Oratory fervently denies this formation's political character even as it continuously privileges and nourishes the needs and desires of select classes, spaces, and races. A rule and regulatory system that is represented as an unwavering reading and response to simple realities boldly defines itself as anti-political. At the foundation of the ploy, then, its language widely creates, paraphrasing Darrel Crilley (1993), a theater of the objective to understand current city realities (i.e. the realness of the frightful new globalism and the need to physically and socially re-make the city one way as a necessary response). It borrows from established understandings of the world (i.e. understandings of people, places, processes) and titillates the public about the rise of a spectacular, ominous reality, new global times and its cast of heroines and villains. Many come to imagine an unfolding spectacle and are left to ponder something ominous: what will happen next.

In this context, organizers against the global trope must use something thorny and rife with pitfalls: the master's tools. That means they must work, at least initially, within an established tool-kit not of their own devising – terms, vocabularies, common imaginings – and make it work for them. It requires, first, toiling at the basic level of contesting vocabularies, meanings, and understandings in common parlance. These vocabularies must be stripped of content and re-coded, but more is required. In particular, people must also be mobilized and effective engagement strategies devised. Discourses have to activate people and implement strategies that can change the actions of growth machines (e.g. discourage the intention of developers to gentrify a neighborhood, compel more investment in poor African-American communities, induce the state to build or protect more affordable housing). Indeed, this chapter has documented successes in this endeavor, as the global trope is increasingly challenged and turned aside.

Yet successful formal resistance, in the final analysis, requires a crucial ingredient, which brings us back to the chapter's beginning: successful tactical resistance (i.e. re-coding acts, processes, and beings in ritualized life). Out of this building block, formal resistance arises. In this sense, the smallest and most individual of acts (e.g. re-imagining ghetto spaces, re-configuring racial and ethnic categories, re-constituting sense of self) create possibilities for more fundamental change. In this context, the key prize here is the creation of something essential, lived spaces that nurture and nourish human life, in more humane and supportive cities. As Mitchell puts it, spaces of decent living is the aim and the prize of urban-transformative movements. To paraphrase Henri Lefebure (1991): "a revolution" like the one to resist the global trope "that does not produce a new space has not realized its full potential; indeed it has failed in that it has not changed life itself . . ." Decent spaces and decent lives, here, are keenly intertwined.

7 The crisis of the rust belt black ghetto

A crisis, today, looms in impoverished Black neighborhoods in America's Rust Belt. These communities have recently devolved into a new kind of social space – "glocal ghettos" – with their stepped-up usage as an apparatus for the flagrant confining and marginalizing of their populations. One key trigger has been an unleashed, imagined global reality, that pushes public policy and planning to disavow the needs of these spaces and people and to isolate them in the name of a civic survival strategy. The public, like before, begrudgingly accepts integration in abstract principle, but now sanctions a necessary social and symbolic gulf with these African Americans that is institutionalized in policy and planning matters. This drive, relentless and agile, effectively offers a sense of an "isolationist expediency" for a continuously invoked source: the common good. While struggles against this global-speak continue, identification of tainted and problematic neighborhoods and people in need of management and control widely colonizes the realm of commonsense.

Residents in these communities have been devastated by this. Growing numbers of desperately poor and dispirited people now struggle to make lives within new institutional and economic circumstances: trapping Workfare programs, low-wage dead-end jobs, church rooted assistance centers, acres of disinvested blocks, predatory retail configurations, weakened health-care providers, and punitive police and court systems. At the same time, the key organization to traditionally assist and enable this youth, public schools, have deteriorated. They, with eviscerated funds and depleted staffs, increasingly operate as institutions of confinement that less educate than hold and contain. Schools struggle to teach this youth within punishing, segregative neoliberal times. The rhetoric of Ronald Reagan and countless Mayors set the wheels in motion after 1980; even they would have a hard time imagining what has now developed.

In total, something vital to quality-of-life has eroded that will be difficult to recover: the sense of community as a psychic safe-haven for people. Even in the Reagan 1980s, with the shrill rhetorical and material assault on these communities, they still contained formal institutions that nurtured identities and collective sustenance (Wacquant 2002). But this has eroded with the annihilation and replacement of indigenous institutions and supports: manufacturing jobs by exploitive dead-end service work, local shops by chain convenience stores, community newspapers by *USA Today*, CETA jobs by Workfare, social service agencies by churches, local taverns by fast-food restaurants, and diverse economies by parasitic economies. Distant and punitive institutions, at every turn, have arisen. Not surprisingly, both material survival and resistance to oppression now widely involve the realm of the informal (i.e. participation in the shadow economy and immersion in new ways of thinking by the making and consumption of music, body performance, and modes of communicative practice).

Yet a point of refinement as we seek to understand this community: these are not identical, one-size-fits-all monolithic entities. These communities vary, evolving in response to distinctive political cultures, economic bases, and institutional frameworks. They also evolve iteratively within contingent histories; ideologies and strategies that sculpt these places ultimately work through a richly textured local (while they ceaselessly collide with processes at broader spatial scales). Thus, opportunity structures and "insidedness" vary in communities like Hough, Bedford-Stuyvesant, and Fairview. In these places, people are cleaved to locally-rooted histories, experiences, and contingently struggled-for opportunity networks. As Routledge (1993) notes, "each locale produces its own set of circumstances, constraints, and opportunities." This book recognizes this, but has seized upon and reported their dominant and generalizable contours: this was deliberate. A belief has underpinned this project: that bringing out the key attributes of an emergent socio-spatial reality, as opposed to cataloging the subtle gradations of all of these spaces in an all-capturing cartographic fix, is important.

A NEW UNEVEN DEVELOPMENT?

However, this global trope and its marginalizing of black ghettos have deeper tentacles that must be identified to fully understand this process. It is currently driven by a broader force, a societally driven uneven metropolitan development across rust belt cities, which sears "the ground" in these cities. This production of uneven development, of course, has always characterized these cities (rooted in a national framework of how places raise revenues and how land and housing markets operate as vehicles for capital accumulation). As Fainstein and Fainstein (1985) note, cities like these have persisted as places of vacillating but periodically wondrous opportunity for profitable fixed investments and state revenue extraction (via the local property tax). The incentive to physically and socially engineer these cities as class balkanized, sets of enclosures has never waned.

David Harvey (1981) excavated the specifics of this balkanizing years ago. He identified the whirl of structurally driven organizations – Realtors, banks, planning agencies, speculation companies – that routinely build and protect housing and real-estate submarkets. This group, as a collective, partitions urban spaces into islands of submarkets that make accumulation and, indeed, "hyper-accumulation" possible. Such making and segmentation of cities flow out of planning specifications, allocation of housing credit, distribution of information about housing vacancies, patterns of land speculation, and City Hall edicts. In this context, many iterations of uneven development have unfolded across these rust belt landscapes to produce islands of affluence, middle-class neighborhoods, interstitial buffering zones, swaths of parkland and open space, and tracts of disinvested and deteriorating neighborhoods.

But I believe rust belt cities are currently experiencing a new kind of ghetto-punishing uneven development, one that is distinctive at multiple levels. At the most observable level, a new rhetorical trigger drives this within the historical specifics of neoliberal ascendancy, the global trope. Its content is multi-textured: within a specified frame of new global times and heightened market imperatives, a mix of new metaphors, metonyms, and signifiers restructures an established script of good and bad guys to make and partition space. The rhetoric extols the virtues of middle and upper-middle class values and "traditions" (nothing new), but in new ways batters the morals, values, and ways of the black poor. This population, in a new arsenal of attack, is now widely cast as consumptive-dysfunctional, productively inept, and shadow residents whose deservedness as citizens is more dubious than before. In this sense, the initial neoliberal oratory on this issue, Reagan's portrayal of "a people" and "a community" falling into cultural dysfunction and locked into an ongoing downward spiral, has been eclipsed.

At the same time, the main source of this rhetoric that directly speaks about and afflicts these African American neighborhoods appears to have shifted. In the 1980s, Reagan spearheaded a revamped national rhetoric that excoriated these spaces and populations. The federal pulpit became synonymous with statements and proclamations about problematic poor black communities. Now, this kind of rhetoric most forcefully flows from the local level. In Indianapolis, Cleveland, Chicago, Pittsburgh, and the like, local editorials, political proclamations, planning assessments, and media reportage more frequently speak this rhetoric (i.e. of now lost and pathological ghetto people and spaces). Thus, local growth machines are, more than ever, determined not to cede public spaces, downtowns, and "stable" neighborhoods to this ghetto population. At stake, to them, is something crucial: a cultural and class re-seizing of the city and tremendous profits in fixed, built investments in an era of hyper real-estate accumulation.

Yet beneath discussion of supposed brute truths about black ghettos is a complex and multi-textured rhetoric. An offered simple, transparent reality conceals an elaborately staged offering, a kind of theater in-the-making. The rhetoric, in its most strident form, "speaks" directly to a specific issue, cancerous ghetto populations and spaces, but is supported by an offered world of people (e.g. roaming kids, distraught single-mothers, savvy street people), places (e.g. new gentrified spaces, "the streets," ghetto parks) and processes (e.g. roaming gangs, aberrant culture). This "support world" stages this assertion's veracity as it connects to the vibrancy of other rhetorics (e.g. prevailing discourses on crime, welfare, and inner city decline) as a supporting measure, what Bakhtin (1981) calls an ongoing intertextuality. This notion of black ghetto realities ultimately gains credibility via a time-tested measure: by discursively entrapping people and taking them inside a constructed world of self-validating truths. People are casually drawn into a world and metaphorically pushed through an assemblage of intuitively resonant images and meanings. In this world one only sees tautological icons that serve as a series of proofs. This world, in the final analysis, reflects the politics and strategies of its makers.

At the same time, this new uneven development unfolds and punishes these black ghettos via a new institutional formation Roger Kiel's (2002) neoliberal amalgam. In form, institutional bases now more profoundly reflect the tenets of neoclassical economic fundamentalism, the prowess of markets, and punitive social strategies. These institutional bases, currently, tilt the value judgments of many in a key terrain, the physical and social restructuring of poor people's worlds. What is being destroyed in these cities – low-income black communities, the welfare apparatus, physical infrastructure, low-cost housing – is being defined and widely understood as "creative destruction" (i.e. the private market being constructively re-asserted to impose efficiency, entrepreneurialism, and hard global realities). Institutional frameworks operate affected by central themes: a need to create a utopia of free markets to guide city dynamics, the necessity of enabling all individuals by unleashing them as entrepreneurial subjects, and a need to privatize the provision of goods and services.

In this setting, neoliberal institutions now supplant and cast in the shadows "welfarist" organizations in rust belt cities. The local state, a key actor, intervenes in local affairs in new ways (e.g. demands waged work for social welfare benefits, privatizes the provision of health and human services benefits) while paradoxically presenting itself as a retrenched local influence. Local assistance agencies, in concert with this, reflect the new neoliberal times and engage the black poor around a pervasive notion: they are a civically unhelpful and maladapted population. Thus, the poor's ability to obtain what were once called entitlements – food stamps, social service help, welfare payments, quality access to public spaces – now requires the likes of engaging faith-based institutions, participating in waged work of any kind, and a proper acknowledgment of one's supposed troublesome values. The poor are treated as troubled but individualized and active subjects that, conditioned by neoliberal tenets, can enhance their own well-being; government now supposedly recognizes this and acts accordingly.

In this context, the historic projection of an all-friendly capitalist rust belt city has turned dour. Once, offers of a sturdy, hard-working place dominated the iconography of rust belt discourses. A gaze onto America presented modest worries and an impulse to aggressively produce commodities and ensure suitable qualities-of-life. Now, the face has turned grim, with talk of a new, harsh competition from distant places and persistent problems from within (e.g. the continuance of enigmatic black ghettos). In this setting, politicians and businesses press for more place-based entrepreneurializing and economic efficiency. In the process, new socially and spatially disciplining policies and programs (e.g. City Growth Plans, tax abatement programs, deepened Workfare, No Child Left Behind, Earnfare) deepen the fragmentation of rust belt cities into the haves and have nots. For the most impoverished African American neighborhoods, the results have been predictable: intense poverty has turned into flagrant deprivation. The ebb and flow of turbulent urban economies, once again, ground these spaces and populations in new social relations and webs of management and control which for the moment appear gripping.

Notes

1. I use the term "black ghetto" in this study with some reservation. The term has a long and complex usage in western academics, and has a deep historical tie with the notion of self-created, socially and culturally devastated terrains. More recently, however, the term has been resurrected with more progressive meanings attached to it (i.e. as spaces of neglect that are complex manifestations of societal processes). I agree in general sentiment with the latter; it is in this vein that I make use of the concept in this work.
2. The irony is that this trend has occurred amid a post-1990 decrease across America in the number of "intensive poverty" census tracts and their agglomerating together in space (see Kingsley and Pettit 2003; Jargowsky and Young 2006). Yet, as post-1990 programs like "Revamped Public Housing" and Hope VI Housing disperse poverty households across wider areas, and deepening gentrification displaces and re-distributes thousands of people, this is anything but surprising. This said, the issue at hand is change in the living and material conditions of this poor who remain in these impoverished African-American neighborhoods.
3. The terms neoliberalism, growth machine, and real-estate capital appear throughout the text. Neoliberalism references a post-1980 form of city governance that more profoundly prioritizes (in rhetoric, programs, and policies) the capabilities of the individual, limited government, and the politics of attracting resources rather than the politics of redistributing resources. The growth machine concept references the constellation of local agencies and institutions (i.e. prominent builders, developers, government, Realtors, the media, local utilities) that form a coalition to work through public policy to enhance "exchange values" in land and property. Real-estate capital, a subset of the growth machine apparatus, is seen to comprise the powerful and traditionally privileged shakers-and-movers in land and property markets (builders, developers, Realtors).
4. A slightly different form of this section and the one that follows ("The Reagan 1980s") was published in David Wilson 2005, *Inventing Black-On-Black Violence: Discourse, Space, Representation* (Syracuse: Syracuse University). It appears here with permission from the publisher.
5. An exception to this 1980s common practice of representing these impoverished black neighborhoods and their populations as falling into a state of cultural and social decline was how these elements were represented on two issues: urban crime and violence (see Hadjor 1993; Wilson 2005). Here caricature and staging had these communities and populations destroyed and dysfunctional. I believe that these distinctive depictions were an outgrowth of conservatives targeting these issues to make their case for something they obsessed with at this time: "the culture wars" and the supposed decline of morality. No issue, according to Eric Schlosser (1998), was more important to conservatives in the 1980s.
6. A slightly different form of this section was published in David Wilson 2005. *Inventing Black-On-Black Violence: Discourse, Space, Representation* (Syracuse: Syracuse University). It appears here with permission from the publisher.

References

Absey, Julie, Darmstadler, Beth and Marcia Egbert (2004) "Three funders: faith based organizations addressing AIDS," *Foundation*, 45(2): 1–2.

ABT Associates (2001) "Turning the corner: Delaware's a better chance Welfare Reform Program at four years," January, Technical report, available from institution.

Adelman, Murray (1988) *Constructing the Political Spectacle*. Chicago: University of Chicago.

Alderman, Michael, Johnson, Michelle, et al. (2002) "Community outreach to reduce disparities in cardiovascular and diabetes morbidity and mortality in the South Bronx," unpublished report, Montefiore Medical Center and Albert Einstein College of Medicine, available from authors of report.

American Atheist (2001) "Philadelphia faith-based director charged with theft but mayor expands controversial program," 17 March, p. 1.

Americans for Democratic Action (1999) "Income and inequality: 8 years of prosperity: millions left behind," technical report, available from institution.

Amin, Ash (2002) "Spatialities of globalization," *Environment and Planning A* 34: 385–399.

Anyon, Jean (1997) *Ghetto Schooling: A Political Economy of Urban Educational Reform*. New York: Columbia University.

Associated Press and Local Wire (2005) "Officials considering state's first private prison," Monday, 21 March, p. 1.

Aziz, Nikhil (2001) "Colorblind-White-Washing America," PRA PublicEye.org, 4 pgs.

Bailey, Nellie (2004) "Forum on growing crisis of black joblessness in Harlem and New York City," website http://educateyourself.org/cn/antiwarmarch17octinwashDC14sep04.shtm

Bakhtin, Mikhael (1981) *The Dialogic Imagination*. Austin: University of Texas.

Balibar, Etienne and Wallerstein, Immanual (1994) *Race, Nation, Class: Ambiguous Identities*. London: Verso.

Balkin, Steve (2002) Discussion with economist and activist, Roosevelt University, 19 June.

Bassett, Keith and Short, John (1981) *Housing and Residential Structure*. London: Routledge.

Bean, F. (2004) discussion with Planner, City of Philadelphia, 12 July.

Beauregard, Robert A. (1993) *Voices of Decline*. Oxford: Blackwell.

Beckett, Katheryne (1997) *Making Crime Pay*. New York: Oxford University Press.

Bein, T. (2004) Discussion with community activist, City of Philadelphia, 1 September.

Berg, Nancy (2003) "Childhood lead poisoning named top environmental threat," technical report, National Lead Watch, Fairfield, Iowa.

Berger, Renee (1997) "People, power, politics". Washington: American Planning Association.

Beveridge, Andrew (2003) "The poor in New York City," *GothamGazette.com*, 3 pgs. Website http:/wwwgothamgazette.com/article/demographics . . . mailasurams.edu/pepermail/asu-sowk-majors/2003-April/000295.html

Beverly, Devira (1991) Discussion with activist, Chicago, 14 July.

Bleifuss, Joel (2002) "No child left unrecruited," *In These Times*, 6 December, p. 4.

Blue Ribbon Panel Report (2002) "Black infant mortality report," New Jersey Health and Senior Services, Trenton, technical report, available from authors.

Boland, Tom (1999) "Homeless protest police sweeps in Cleveland, Ohio, USA," website http:aspin.asu/hpri/archives/Dec99/0174/html

Boyer, Christine (1983) *Dreaming the Rational City*. Cambridge: MIT.

Brady, David (2005) Commentary in "75th Anniversary celebration: Panel on an ethic for a new global economy." Website http://www.sb.stanford.edu/news/headlines/globalethic.shtml

Braggs, B. (2004) Discussion with Planner, Department of Metropolitan Development, City of Indianapolis, 2 August.

Brenner, Neil, Jessop, Bob, Jones, Martin and MacLeod, Gordon (2003) *State/space: a reader*. Oxford: Blackwell.

Brenner, Neil and Theodore, Nik (2002) "Cities and the geographies of actually existing neoliberalism." In N. Brenner and N. Theodore (eds), *Spaces of Neoliberalism: Urban Restructuring in North America and Western Europe*. Oxford: Blackwell, 1–33.

Brookings Institution (2004) "Event summary: block grants, past, present, and prospects," *Cities and Suburbs Website*, 15 October. Website www.brookings.edu/metro/speeches.htm?show=all

Brookings Institution (2005) "The price is wrong: getting the market right for working families in Philadelphia." Technical report, 1755 Massachusetts Avenue, Washington D.C., available from institution.

Brown, Matt (2003) "The problem of school funding," December, *The Plain Press*, p. 1.

Buck, Marilyn (2004) "The U.S. Prison System," *Monthly Review*, 55: 6–13.

Burnier, DeLysa and Descuter, David (1992) "The city as marketplace: a rhetorical analysis of the urban enterprise zone policy." In A. Weiler and W. B. Pierce (eds), *Reagan and Public Discourse in America*. Tuscaloosa: University of Alabama, pp. 251–265.

Bush, George, W. (2002) "Forward," White House position paper, available from Government Printing Office, Washington D.C.

Bush, George (2004) Speech "On the war on terror," transcript from presentation at Roswell Convention and Civic Center, Roswell, New Mexico, January.

Business Week (1979) "Smaller cities, with no end to suburbanization," September 3, pp. 204–209.

Butterfield, Greg (1998) "Odor of New Workfare Boss in NYC Proceeds Him," *Workers World*, 1, 3–5.

Butterfield, Patricia and Larrson, Laura (2003) "Protect children from lead exposure," *Discovery*, May: 2–3.

Cameron, R. and Palen, R. (2003) "The imagined economy: mapping transformation in the contemporary state." In N. Brenner, B. Jessop, M. Jones and G. Macleod (eds), *State/space: a reader*. Oxford: Blackwell, pp. 165–184.

Campbell, Jane (2003) "State of the city, text from technical report, unpublished document.

Campbell, Jane (2004) "State of the city," speech from mayor of Cleveland, text from technical report, unpublished document.

Carley, David (1987) Discussion with Head, Department of Metropolitan Development, City of Indianapolis, July 19.

Carnegie Mellon Alumni Magazine (2002) "The new entrepreneurs," Spring, 2–5.

Caro, Robert (1974) *The Power Broker: Robert Moses and the Fall of New York*. New York: Vintage.

Carpenter, Dave (2003) "Manufacturing jobs vanish as Illinois economy evolves," technical report, Kellogg School of Management, Northwestern University.

Castells, Manuel (2004) *The Power of Identity*. Oxford: Blackwell.

Center on Budget and Policy Priorities (2002) "Oversight Hearing on the Federal Deposit Insurance System and Recommendations for Reform," prepared statement before the U.S. Senate Committee on Banking, Housing, and Urban Affairs, 1 May, transcript available from Committee.

Center for Economic Development (2004) "After the boom: joblessness in Milwaukee," technical report, available from Center, Milwaukee, Wisconsin, 14 pgs.

Center for Juvenile and Criminal Justice (2002) "Seeking Balance: Reducing Prison Costs in Times of Austerity." Technical Report, available from institution.

Center for Research On Religion (2002) "Then comes marriage? Religion, race, and marriage in urban America", technical report, available from Center.

Center on Urban Policy and Social Change (2001) "The end of welfare as they knew it: what happens when welfare recipients reach their time limits?" Technical report, Mandel School of Applied Social Sciences, Case Western Reserve University.

Center on Urban Policy and Social Change (2004) "Testing measures of neighborhood change: the Cleveland community building initiative (CCDI)," Technical report, Mandel School of Applied Social Sciences, Case Western Reserve University.

Chandler, Susan (2004) "Gangs built on corporate mentality," *Chicago Tribune*, 13: June 3.

Cheyfitz, Eric (1981) *The Poetics of Imperialism*. New York: Oxford University Press.

Chicago Gang Research Project (2003) "For Richard J. Daley to Larry Hoover: from Al Capone to Gino Colon," technical report, available from Project, Chicago.

Chicago Reporter (2000) "Drug arrests nab minorities," 2 March, p. 2.

Chicago Sun-Times (1996) "New homes key to Woodlawn redevelopment," 17 December, p. 14.

Chicago Tribune (1994) "Neighbors take back the streets," 18 September: 4.

Chicago Tribune (2003) "Gangs run pipeline from Delta to Chicago," 13 November, p. 9.

Chicago Tribune (2004) "Chicago surveillance cameras to be fitted with listening devices," 7 April, p. 12.

Chicago Tribune (2004a) "The Working Poor," 25 April, Sunday, Section 1, pp. 1–6.

Child and Family Health Services (2004) "Child and Family Health Services Community Health Indicators Project," Cuyahoga County Board of Health, City of Cleveland, technical report.

CIA World Fact Book (2003) Technical report, federal government document, Washington D.C.
City Club of Cleveland (2006) "An Entrepreneurial Approach to Improving Urban Education," 22 March, transcript from City Club.
City of Indianapolis (2003) "The Consolidated Plan," technical report, Division of Community Development, Department of Metropolitan Development, available from City.
City of Philadelphia (2004) "Community Report Card: Upper North." Technical report prepared by City of Philadelphia, available from City.
Civic Task Force on International Cleveland (2003) "Recommendations to the City of Cleveland," technical report, City of Cleveland, available from City.
Clarke, N. (2004) Discussion with community organizer, Chicago, 21 July.
Cleveland City (2004) "Department of Community Development – Block Grant Fact Sheet," website www.city.cleveland.oh.us/government/departments/commdev/cdblockgrantprog.html
Cleveland Indy Media Center (2004) "Friday, June 11: Tremont Anti-Gentrification Rally," website Cleveland.indymedia.org/news/2004/06/10988.php
Cleveland Plain Dealer (2005) "Cosby gets it wrong, 14 May, p. B-1.
Cleveland Plain Dealer (2005a) "Guiding goals along a more promising path," 14 April.
Cleveland Social Services (2004) Discussion with Head, 7 September.
Cleveland State University (2001) "Urban update: news from the Maxine Goodman Levin College of Urban Affairs," Fall: 1–3.
Cline, P. (2004) Quote in "Carol Marin: Gangs, police become more sophisticated," NBC5*.com*, 2 pgs, 20 May. Website www.officer.com/article/article.jsp
Coastal Post Online (2003) "The Draft: No Child Left Behind," website ns/boomframe.jsp
Cochran, Tom (2003) Interview in "U.S. Conference of Mayors Releases New Report Showing That Economy Surges, But With Lower Paying Jobs," press release, U.S. Conference of Mayors, Monday 10 November.
Collins, Sheila (1996) *Let Them Eat Ketchup*. New York: Monthly Review.
Columbus Post (2003) "One officer suspended, another resigns in Profiling at Mall," 18 September, p. 10.
Common Dreams News Wire (2006) "Alito Rally Pastor received 'Faith Based' funds from Bush Administration, Watchdog Group Reports," 4 January, website http://www.commondreams.org/news2006/0104-03.htm
Community Voices Heard (1999) "Welfare-to-Work: is it working? A case for public jobs creation," technical report, available from organization, New York.
Cox, Kevin (1993) "The local and the global in the new urban politics: a critical review," *Environment and Planning D: Society and Space*, 11: 433–448.
Cox, Kevin and Mair, Andrew (1987) "Locality and community in the politics of local economic development," *Annals of the Association of American Geographers*, 88: 307–325.
Crain's Cleveland Business (2005) "Salvation in startups: can Mom-and-Pop shops help the city rise from an economic slump," 14 February, p. 17.
Creswell, Tim (1996) *In Space/Out Space*. Minneapolis: University of Minnesota.
Crilley, Darrel (1993) "Architecture as advertising: constructing the image of redevelopment." In G. Kearns and C. Philo (eds), *Selling Places: The City as Cultural Capital, Past and Present*. Oxford : Pergamon, pp. 231–252.
Cronin, Mike (2004) Quote in "Carol Marin: Gangs, police become more sophisticated," NBC5.com, 2 pgs, 20 May.
Daily Iowan (2005) "Iowa faces prison woes," 2 December, p. 1.

Davis, Mike (1990) *City of Quartz*. London: Verso.

Davis, Sue (2002) "Harlem rally slams rent hikes," *Workers World*, 18 July: 12.

Dear, Michael (2000) *The Postmodern Urban Condition*. Oxford: Blackwell.

Dear, Michael (2002) *From Chicago to L.A: Making Sense of Urban Theory*. Newbury Park: Sage.

Dear, Michael and Clark, Gordon (1984) *State Apparatus*. Boston: Allen & Unwin.

Dear, Michael and Flusty, Steven (2001) *The Spaces of Postmodernity: Readings in Human Geography*. Oxford: Blackwell.

De Certeau, Michel (1984) *The Practice of Everyday Life*. New York: Basic.

Defilippis, James (2004) *Unmaking Goliath*. London: Routledge.

Delaware Valley Regional Planning Commission (2002) "Regional data bulletin," technical report, #74, October.

DeMause, Neil (2002) "Bad to worse: welfare reform is up for reauthorization, but it's only going to get meaner," *In These Times*, 2 August: 7.

Democracy Now (2005) "On the military," presentation, WEFT radio, Urbana-Champaign, Illinois, 3 March.

Department of Housing and Urban Development (2002) "2002 Annual Report," technical document prepared by the U.S. Department of Housing and Urban Development, available from institution.

Detroit Evening News (2001) "Schools ignore vacant houses," 25 November, p. 14.

Diamond, Sarah (1995) *Roads to Domination: Right Wing Movements and Political Power in the United States*. New York: Guilford.

Dicken, Peter and Lloyd, Peter E. (1990) *Location in Space*. New York: Harper and Row.

DiMora, Jimmy (2005) "Commissioners poised to boost Arts, again! Media Release, Cuyahoga County Department of Development, 26 May.

Diwan, Ramesh (1998) "Globalization: myth versus reality," "Essays on the policy of Swadeschi," technical report, Department of Economics, Rensalaer Polytechnic Institute, New York, available from author.

Drake, St. Clair and Cayton, Horace (1945) *Black Metropolis*. Chicago: University of Chicago.

D'Souza, Dinesh (1999) Talk at University of Illinois at Chicago, February.

Eagleton, Terry (1991) *Ideology*. London: Verso.

Economist.com (2003) "America's widening rich-poor gap," 4 September. Website www.officer.com/article/article.jsp

Education Partnership Program (2004) "Conservative Workfare," website www.datacenter.org/programs/econ_justarchive03.htm

Egan, Paul (2003) "Tenants hold no hope for repairs," *Detroit News*, 17 June.

Ehrenreich, Barbera (2002) "Nickeled and dimed," *Now With Bill Moyers*, 29 March, PBS movie and transcript, available from National Public Broadcasting, Washington.

Elliot, S. (2004) Discussion with Planner, City of Cleveland, 13 August.

Epps, B. (2004) Discussion with Planner, City of Cleveland, 10 April.

Erie, G. (2001) "Erie receives award, national press release, National League of Cities, website www.cnn.com/2001/LAW/03/12/ashcroft.profiling/

Ervin, Lorenzo K. (2000) "It's racism stupid," *NewsKiosk*, 1 September, p. 1.

Eviatar, Daphne (2001) "Murdoch's Fox News," *Nation*, 22 February.

Fainstein, S. S. and Fainstein, N. (1985) "Economic restructuring and the rise of urban social movements," *Urban Affairs Quarterly*, 21: 187–206.

Fairclaugh, Norman (1992) *Discourse and Social Change*. New York: Polity.

Farella, Carrie (2003) "Breath savers – RNs attack asthma at Chicago's Ground Zero," *Nursing Spectrum*, 13 July, 2–14.
Feagin, Joe (1979) "Excluding blacks and others from housing: the foundations of white racism," *Cityscape*, 4: 78–90.
Feldman, Linda (2003) "Faith-based initiatives quietly lunge forward," *Christian Science Monitor*, 6 February, p. 10.
Ferman, Barbara (1996) *Challenging the Growth Machine in Chicago*. Lawrence: University of Kansas.
Fertig, C. (2004) Discussion with Planner, City of Chicago, 18 August.
Field, M. (2004) Discussion with Planner, City of St. Louis, 24 August.
Fingleton, Enmonn (1999) "In praise of hard industries," *IndustryWeek.Com* www.industryweek.com/currentArticles/articles.asp?
First Amendment Center (2004) "Chicago police will enforce new curfew ordinance," www.firstamendmentcenter.org, 21 March.
Fisher, Eric O. (2004) "Why are we losing manufacturing jobs," Federal Reserve Bank of Cleveland, technical report, July.
Fitrakis, Bob (2000) "Racial Profiling in Columbus: Is It A Crime to Drive, or Fly, While Black?" *The Free Press*, 29 July, p. 8.
Fitzgerald, Thomas (2004) "Bush touts AIDS funds," *Philadelphia Inquirer*, 24 June, p. 16.
Foner, Philip S. (2002) *The Black Panthers Speak*. New York: Da Capo.
Forman, Robert (1971) *Black Ghettos, White Ghettos, and Slums*. Englewood Cliffs: Prentice-Hall.
Foster, J. B. (2002) "Monopoly capital and the new globalization," *Monthly Review*, 53: 18–24.
Franks, Fiscella D., Gold, P. and Clancy, C. M. (2000) "Inequalities in quality: addressing socioeconomic, racial, and ethnic disparities in health care," *Journal of the American Medical Association*, 238: 2579–2594.
Freelance Wire Services (2004) "Chaotic prison system thinks it's above the law," *Hollister Free Lance*, Tuesday, 24 February.
Fried, M. (2004) Discussion with local planner, City of Chicago, 11 August.
Friedman, Lawrence M. (1980) "Public housing and the poor." In J. Pynoos, R. Schafer and C. W. Hartman (eds), *Housing Urban America*. New York: Aldine, pp. 473–484.
Friedman, Milton (1995) "Public schools: make them private," Briefing Paper No. 23, Cato Institute, available from institution.
Friedman, Thomas L. (1999) *The Lexus and the Olive Tree*. New York: Farmer, Strauss and Giroux.
Gainsborough, Jenni and Maves, Marc (2004) "Diminishing returns: crime and incarceration in the 1990s," technical report, the Sentencing Project, 514 10th Street N.W., Washington D.C.
Garretson, D. (2004) Discussion with local activist, City of Cleveland, 17 August.
Gasping For Justice (2003) No Mas Pollusion! Website www.gaspingforjustice./20fr.com/
Gelfand, Mark (1965) *A Nation of Cities: the Federal Government and Urban America*. New York: Oxford University Press.
Gibson-Graham (2002) "Beyond global versus local: economic politics outside the binary frame." In A. Herod and M. W. Wright (eds), *Geographies of Power: Placing Scale*. Oxford: Blackwell, pp. 25–50.
Gillespie, Rey (1966) *The Anatomy of Riots in Hough*. New York: Harper & Row.

Glazer, Nathan (1970) *Cities in Trouble*. Chicago: Quadrangle.
Goldsmith, Stephen (2003) "Putting faith in neighborhoods: making cities work through grassroots citizenship," Indianapolis, Hudson Institute, technical report.
Goldsmith, Stephen (2003a) "Introduction." In *The Entrepreneurial City: A How-To Handbook For Urban Innovators*. New York Manhattan Institute For Policy Research, pp. 1–10.
Goldstein, Fred (2001) "Police brutality – From L.A. to New York, it happens every day," *Workers World*, 22: 11–17.
Goodrich, Peter (1987) *Legal Discourse: Studies in Linguistics, Rhetoric, and Legal Analysis*. New York: St. Martins.
Gorenstein, Nathan, Boyer, Barbara and Ciotta, Rose (2006) "Shootings Ravage City Neighborhoods," *Philadelphia Inquirer*, 24 May, p. 6.
Gruley, Bryan (2001) "USA: prison building spree creates glut of lockups," *Wall Street Journal*, 6 September, p. 8.
Grunwald, Michael (1998) "The myth of the supermayor," *American Prospect*, 40, 1 September, 4–7.
Habermas, Jurgen (1973) *Legitimation Crisis*. Boston: Beacon.
Hacker, Andrew (1992) *Two Nations*. New York: Scribner.
Hadjor, Kofi Burnos (1993) *Another America*. Boston: South End.
Hamel, Chris (2001) "People's conquers plans speakout of workers and poor," *Workers World*, 28 June: 4.
Harvey, David (1974) "Class-monopoly rent, finance capitals, and the urban revolution," *Regional Studies*, 8: 239–255.
Harvey, David (1981) *The Limits to Capital*. Oxford: Blackwell.
Harvey, David (1985) *The Urbanization of Capital*. Oxford: Blackwell.
Haworth (1966) *The Good City*. Bloomington: Indiana University.
Heilbroner, Robert (1976) *The Worldly Philosophers*. New York: Simon and Schuster.
Helper, Rose (1969) *Racial Policies and Practices of Real Estate Brokers*. Minneapolis: University of Minnesota.
Henderson, Charles (1999) "George Bush invites God to school," *Christianity Newsletter*, 8 November, p. 1.
Hennepin, Bob (2004) Discussion with Planner, City of Cleveland, 4 March.
Herod, Andy and Wright, Melissa (2002) *Geographies of Power: Placing Scale*. Oxford: Blackwell.
Herron, Jerry (1993) *After Culture: Detroit and the Humiliation of History*. Detroit: Wayne State University.
Hirsch, Arnold (1976) *A Ghetto Takes Shape: Black Cleveland, 1870–1930*. Chicago: University of Chicago.
Holding, S. (2004) Discussion with Planner, department of Metropolitan Development, City of Indianapolis, 11 October.
Hollway, Wendy (1984) "Gender difference and the production of subjectivity." In J. Henriques, W. Hollway et al. *Changing the Subject: Psychology, Social Regulation, and Subjectivity*. New York: Metheun.
Holston, James and Apparadurai, Arjun (2003) "Cities and citizenship." In N. Brenner, B. Jessop, M. Jones and G. MacLeod (eds), *State/Space: A Reader*. Oxford: Blackwell, pp. 296–308.
hooks, belle (1993) *Outlaw Culture*. London: Routledge.
Hough Neighborhood History (2003) Website www.link.net/spa/houghhist.htm

Houghton Mifflin (1991) *Readers Guide To American History*, Eric Foner and John Garrety (eds).

Howard, Glen (1991) Discussion with City–County Councilor, City of Indianapolis, 14 May.

Humphrey, Hubert (1948) "What's wrong with our cities?" *The American City* 63 (July): 12–13.

Hurley, J. (1995) *Environmental Inequalities*. Lanham, MD: Rowman & Littlefield.

Hurlich, Susan (2001) "The twelth caravan: reversing the challenge," website www.nscuba.org/DocsBloque/Hurlich – 13 – July – 01.html

Illinois Board of Education (2003) "No Child Left Behind (NCLB), technical document, State of Illinois, Springfield, available from Illinois Board of Education.

Illinois Department of Human Services (2002) "Earnfare," technical document, available from agency.

Immergluck, Dan (2005) "The power of a community based development corporation," *Shelterforce*, 141, 15 May: 4–7.

Independent Media of Philadelphia (2004) "Alternatives to corporate globalization," 5 January, website www.phillyimc.org/alternatives/

Indianapolis Convention and Visitors Association (2003) "Indy," unpublished pamphlet, available from Visitors Association.

Indianapolis Regional Center Plan (2001) "Comeback cities: the prospects for a continuing urban renaissance," technical report, available from City of Indianapolis, Department of Metropolitan Development.

IndyStar.Com (2001) "Library fact files, Stephen Goldsmith," website www.indystar.com

Innovation Philadelphia (2004) "Philadelphia's love park goes wireless," 16 June, 17, 1, p. 1.

Institute for Civic Infrastructural Systems (2004) "South Bronx environmental health and policy study," technical report, New York University, New York City, available from authors.

Irons, John S. (2003) "2001 recession in perspective: economic and budget situation, Technical report, Office of Management and Budget, Washington D.C.

Jakle, John and Wilson, David (1992) *Derelict Landscapes*. Lanham, MC: Rowman and Littlefield.

Jargowsky, Paul (1997) *Poverty and Place: Ghettos, Barrios, and the American City*. New York: Russell Sage Foundation.

Jargowsky, Paul (2002) "Stunning progress, hidden problems: the dramatic decline of concentrated poverty in the 1990s." Technical Report, the Brookings Institute, Washington D.C.

Jargowsky, Paul and Young, Rebecca (2006) "The Underclass Revisited: A Social Problem in Decline," *Journal of Urban Affairs*, 28: 55–70.

Jenks, Chris (2004) Discussion with Planner, City of Cleveland, 15 August.

Jessop, Bob (1990) *State Theory: Putting the Capitalist State In Its Place*. Oxford: Polity.

Jet Magazine (1995) "Riots, unrest plague three cities after police shootings, allegations of brutality," 21 August, p. 11.

Johnson, David (2003) "Important cities in black history," *Feature*, website www.factmonster.com/spot/bhmcities1.html

Johnston, William, Jr. (2005) "Our present," text of speech delivered at the New York State Network For Economic Research, 30 March, available from City of Rochester.

Jonas, Andrew E. G. and Wilson, David (1999) *Two Decades Later: Critical Perspectives On the Growth Machine Thesis*. Albany: SUNY.

Judd, Dennis (1979) *The Politics of America's Cities: Private Power and Public Policy*. Glenville, Ill: Little and Brown.

Katz, Bruce and Jackson, David (2004) "Purging the parasitic economy," Cities and Suburbs technical report, the Brookings Institution, 7 September, 7 pgs.

Keil, Roger (2002) "Common-Sense" Neoliberalism: Progressive Conservative Urbanism in Toronto, Canada." In N. Brenner and N. Theodore (eds), *Spaces of Neoliberalism: Urban Restructuring in North America and Western Europe*. Oxford: Blackwell, 230–253.

Keith, Michael (1993) *Race, Riots, and Policing*. London: UCL Press.

Keller, Michele (2002) "Bush proposal would pay Workfare recipients less than minimum wage," technical report, National Association of Women, available from Association.

Kelley, Robin (1997) *Yo Mama's Disfunctional*. New York: St. Martins.

Kelly, Fred (2003) "Ministers to wage anti violence campaign," *Indy News*, 11 October, p. B-1.

Kho, Byron (2004) "Mayor plans to overlay city with wireless internet access," *The Daily Pennsylvanian*, 8 September, p. 1.

Kilpatrick, J. (2002) "Mayor Kilpatrick's Inaugural Speech," City of Detroit, transcript available from City.

Kingsley, G. Thomas and Pettit, Kathryn L. S. (2003) "Concentrated Poverty: A Change In Course." In *Neighborhood Change in Urban America*, #2. Washington D.C.: Urban Institute.

Kleiman, Mark (2003) "Faith-based fudging: how a Bush-promoted Christian prison program fakes success by massaging data," technical report, MSN website www.moonfarmer.org/archives/2003_08.php.

Knox, Paul (1997) "Globalization and urban economic change," *Annals of the American Academy of Political and Social Science*, 551: 17–28.

Koch, Kathleen (2001) "Urban America looks to future with an eye on the past," 10 March, transcript available from CNN, Atlanta.

Konkel, Mark J. (2005) "City to shame hookers, johns online," *Chicago Sun-Times*, 22 June, p. 18.

Koolhas, Rem (2003) "Delirious No More," Post to Wired New York Forum, 8 June, http://www.wirednewyork.com/forum/showthread.php?+=3881

Kopel, B. David (2001) "Crime: the inner city crisis," website www.achr.net/ACHR%2015%20with%20photos.pdf

Kropko, M. R. (2000) "Cleveland settles homeless lawsuit with ACLU," *Akron Beacon-Journal*, Wednesday, 2 February, p. 6.

Kusmer, Kenneth L. (1976) *A ghetto takes shape: black Cleveland, 1870–1930*. Chicago and Urbana: University of Illinois.

Lakoff, G. and Johnson, M. (1980) *Metaphors We Live By*. Chicago: University of Chicago.

Lauria, Mickey (1999) "Reconstructing regime theory: regulation and institutional arrangements." In A. E. G. Jonas and D. Wilson (eds), *Two Decades Later: Critical Perspectives On the Growth Machine*. Albany: SUNY, pp. 183–205.

Law, Bob (2000) *From Black Rage To A Blueprint For Change*. Chicago: African American Images.

Laws, Glenda (1997) "Globalization, immigration, and changing social relations in cities," *Annals of the American Academy of Political and Social Sciences*, May, 551, 89–105.
Lefebure, Henri (1991) *The Production of Space*. Oxford: Blackwell. Translated by Donald Nicholson-Smith.
Leile, J. (2004) Discussion with activists, City of Philadelphia, 3 September.
Lemann, Nicholas (1992) *The Promised Land*. New York: Vintage.
Levi, Robin and Appel, Judith (2003) "Collateral consequences: denial of basic social services based upon drug use," technical report for Drug Policy Alliance, 13 June, available from Drug Policy Alliance.
Levitan, Sar (1985) *Programs In Aid of the Poor*. Baltimore: Johns Hopkins.
Lieske, Joel A. (1978) "The conditions of racial violence in American cities," *American Political Science Review*, 72: 1324–1340.
Lilly Endowment (2003) "Lilly Endowment Issues 2003." Annual Report, technical document, available from Endowment.
Lipietz, Alan (1994) "The national and regional: their autonomy vis-à-vis the capitalist world crisis." In R. Palen and B. Gill (eds), *Transcending the State-Global Divide*. Boulder: Riener, pp. 23–44.
Lipman, Pauline (2005) "We're not blind: just follow the dollar sign," *Rethinking Schools*, 19(4) Summer: 1–6.
Livingstone, David (1992) *The Geographical Tradition: Episodes in the History of A Contested Enterprise*. Oxford: Blackwell.
Lloyd, B. (2004) Discussion with Workfare worker, City of Chicago, 11 August.
Long, Robert E. (1989) "The Workfare debate," *Muckraker* http://www.muckraker.org/inv_by_topic.php?topic_id=7&&show_all=1'
Lott, Tommy (1999) *The Invention of Race*. Oxford: Blackwell.
Loughner, Jamie (2002) "U.S. Washington D.C. Housing Occupation – open and shut case," website www.ainfus.ca/02/aug/ainfos00134.html
Loughner, Jamie (2004) Discussion with activist, Washington D.C., 22 November.
Loury, Glen (1996) "The divided society and the democratic idea," transcript of University Lecture, Boston University, 7 October.
MacLeod, Gordon (2002) "From urban entrepreneurialism to a revanchist city? On the spatial injustices of Glasgow's Renaissance." In N. Brenner and N. Theodore (eds), *Spaces of Neoliberalism: Urban Restructuring in North America and Western Europe*. Oxford: Blackwell, pp. 254–277.
Maharidge, Dale (2004) "Rust and rage in the heartland," *Nation*, 20 September, 3–11.
Maharoj, Nicole (2001) "Mayors hold first national forum on faith-based and community initiatives," *U.S. Mayor Articles*, 9 July, technical report.
Maharaj, Nicole and Bullock, Derrick (2003) "Mayoral leadership on faith-based and community initiatives workshop," *U.S. Mayor Articles*, technical report.
Maher, Tim (2003) Discussion with Professor, University of Indianapolis, 23 June.
Mail Archive (2002) "A problem with Workfare," Saturday, 28 March, website mail-archive.com/futurework@dijkstra.uwaterloo.ca/msgo3102.htm.
Malanga, Steve (2003) "Downtown rebuilding gets serious," *City*, 25 April, 1–4.
Marable, Manning (1997) "Full employment and affirmative action," technical report, National Jobs For All Coalition, available from Coalition, Washington D.C.
Marcuse, Peter (1978) "Housing policy and the myth of the benevolent state." In *Critical Perspectives on Housing*, R. Bratt, C. Hartman and A. Meyerson (eds). Philadelphia: Temple University, pp. 77–93.

Martin, Ron (1994) "Economic theory and human geography." In D. Gregory, R. Martin and G. Smith (eds), *Human Geography: Society, Space, and Social Science*. Minneapolis: University of Minnesota.

Martindale, Scott (2005) "Army personnel are putting in long hours," *Daily Breeze*, website www.mothersagainstthedraft.org/index

Masotti, Louis H. (1968) "Toward an escalation of civil violence: the Glenville (Cleveland) incident, July 23–28, 1968," technical report, Case Western Reserve University.

Massey, Doreen (1999) "Space-time, 'Science' and the relationship between physical geography and human geography. *Transactions of the Institute of British Geographers* 24: 261–276.

Massey, Douglas S. and Denton, Nancy A. (1994) *American Apartheid*. Boston: Harvard University.

Matthews, M. B. (2005) "Street Smarts," website http://mbmatthews.blogsprt.com/2005/10/Cleveland – third – world – city.html

McCarthy, Cameron (1995) "Danger in the safety zone: notes on race, resentment, and the discourse of crime, violence, and suburban security." University of Illinois, Department of Education Policy Studies, Urbana, unpublished manuscript.

McDowell, Linda (1995) "Understanding diversity: the problem of/for theory." In R. J. Johnston, P. J. Taylor and M. J. Watts (eds), *Geographies of Global Change*. Oxford: Blackwell, pp. 280–294.

McGrath, N. (2003) Discussion with City Councilperson, City of Chicago, 11 July.

McKenzie, James (2003) "The cheapening of America: NCLB and the decline of the good job," *NoChildLeft*, 1, 12, December.

McKenzie, James (2003a) "Gambling with the children," *NoChildLeft*, January, 2, 1, January.

McKinney, Frank (2002) *Make It Big: 49 Secrets for Building A Life of Extreme Success*. New York: John Wiley.

McLellan, Thomas A. (2003) "Cutting addictions treatment: what does the science tell us?" Technical document, available from Treatment Research Institute, Philadelphia.

McWhorter, John (2002) "Job loss didn't make the underclass," *City Journal*, Autumn, 12: 2–4.

Media Ventures (1999) "Pacers open new fieldhouse," technical report, Indianapolis, November, available from authors of document.

Melosi, Martin (1981) *Garbage in the Cities: Refuse, Reform, and the Environment 1880–1980*, College Station: Texas A + M.

Mercer, Kobeena (1997) *Welcome to the Jungle*. London: Routledge.

Midwest Partners (2003) "State of the States: tough fiscal choices, technical report, website www.lwvil.org/download.php?file_name=status_of_the_state_2002-11.pd

Miller, David (2003) "Unspinning the globe," *Red Pepper*, 6–8 June.

Miller, Zane (1973) *The Urbanization of Modern America*. New York: Penguin.

Miner, Barbara (2004) "Keeping public schools public: testing companies mine for gold," *RethinkingSchoolsOnline*, Winter, website www.rethinkingschools.org/special.reports/bushplan/index.html

Mitchell, Don (2000) *Cultural Geography*. Oxford: Blackwell.

Mokhiber, Russell and Weissman, Robert (2003) "Stealing money from kids," *ZNet*, website www.zmag.org/content/Economy/mokweiss_stealingfromkids.cfm

Morial, Marc (2002) "The new world," speech to the National Urban League, August, transcript available from the Urban League.

Mosiman, Dean (2002) "Mayors get boost from corporations," *Wisconsin State Journal*, 13 June, p. 1.
Muller, Mancow (2001) Radio commentary from the Mancow Muller Morning Madhouse Show, WPGU, Champaign, Illinois.
Muller, Mancow (2002) Radio commentary from the Mancow Muller Morning Madhouse Show, WPGU, Champaign, Illinois.
Mung, D. (2004) Discussion with activist, City of Chicago, 22 November.
Nader, Ralph (1997) "Hold a Baltimore's point of view," *League of Fans*, website www.leagueoffans.org/coldobaltimorecolumn.html
National Center for Health Statistics (2006) "Health of black or African American populations," website http/www.cdc.gov/nhs/fastats/black_health.htm
National Center for Neighborhood Enterprise (2003) "Empowering America's communities," technical report, available from Center, Washington D.C.
National Governors Association (2002) "State fiscal outlook continues to deteriorate," technical report, available from Association.
National Law Center (1994) "Report finds cities are increasingly criminalizing homelessness and poverty," technical document, available from organization.
National Youth Gang Center (1998) "1998 National Youth Gang Survey", technical report, Office of Juvenile Justice and Delinquency Prevention, U.S. Department of Justice.
Nawojczyk, Steve (1997) "Street gang dynamics," technical report, the Nawojczyk, Group, available from organization, 9 pgs.
Naymik, Mark (2003) "Debt service: financial disaster threatens: a major Eastside Community Center," website www.clevescene.com/issues/1990-0.7-08/news.html
New York Public Library Digital Project (2004) "The urban crisis," technical report available from institution, 188 Madison Avenue, New York.
New York Times (1998) "Newark Mayor Sharp James praised for revival of downtown", 27 February, p. 28.
New York Times (2005) "In the ring but attached to the cell block," 16 July, p. 1-D.
Nichols, Laura and Gault, Barbara (1999) "The effects of Welfare Reform on housing stability and homelessness," *Welfare Reform Network News*, 2 March, 1–13.
Noll, P. (1990) *Religion and American politics: approaches and interpretations*. Oxford: Oxford University.
Norman, Michael (1993) "One cop, eight square blocks," *New York Times Magazine*, 12 December, pp. 62–71.
Norquist, John (1998) *The Wealth of Cities*. New York: Addison Wesley Longman.
Norquist, John (2000) "John Norquist address," public address, website www.landmarkcenter.org/norqtext2.html
Norris, Christopher (1982) *Deconstruction: Theory and Practice*. New York: Methuen.
Norton, Anne (1993) *Republic of Signs*. Chicago: University of Chicago.
O'Hare, Madelyn M. (2004) "Philly faith-based director charged with theft," *American Atheist*, website www.atheists.org/visitors.center/OHairFamily
O'Hanian, Susan (2003) "Bush Flunks School," *The Nation*, 1 December, 9–14.
O'Malley, Martin (2001) State of the city," speech in Baltimore, 1 February, available as technical report, City of Baltimore.
Online Insider (1999) "An (Eli) Lilly Blooms in Indianapolis: incentives of $214 Million Power $1 Billion, 7,500-job deal," website http://www.conway.com/ssinsider/incentive/ti9909.htm

Online News Hour (1997) "Is Workfare moral?" *Online Forum*, 6 August.

Oren, Tasha (2001) "Gobbled up and gone: globalization and the preservation of local culture," paper presented at the Global Citizens Conference, University of Wisconsin-Milwaukee, 6 April.

Osofsky, Gilbert (1967) *Harlem: the Making Of A Ghetto*. New York: Harper.

Ostrum, Neenyah (1995) "Are HHF-G +CFS fueling the asthma epidemic in the South Bronx?" *Online News Hour*.

Pablo, Juan (2004) "Hit the Road Sam: Communities Speak Out Against Army's Hip Hop Recruitment," *Village Voice*, 1 March, pp. 24–26.

Pacific Business News (2004) "NBTA warns economic recovery: threatened by travel barriers," 14 May, p. 11.

Padilla, Felix (1992) *The gang as an American enterprise*. New Brunswick: Rutgers University.

Page, Max (2001) *The Creative Destruction of Manhattan, 1900–1940*. Chicago: University of Chicago.

Palast, Greg (2004) "No Child Left Behind: the new educational eugenics in George Bush's State of the Union," *Observer*, Saturday, 24 January, p. 16.

Papachristos, Andrew (2004) "Gangs in the global city: the impact of globalization on post-industrial street gangs," paper presented at the Workshop on the Sociology and Culture of Globalization, Chicago, 3 February.

Parenti, Christian (2000) *Lockdown America: Police and Prisons In the Age of Crisis*. London: Verso.

Parr, John (1998) "Detroit: struggling against history," technical document, Academy of Leadership Group, University of Maryland.

Payne, Fred (2002) "Eminem's real Detroit," *National Review Magazine Online*, 18 November. Website www.**nationalreview**.com/comment/comment-**payne**111802.asp

Peck, Jamie (1995) "Moving and shaking: business elites, state localism, and urban privatism," *Progress in Human Geography*, 19: 16–46.

Peck, Jamie (2001) *Workfare States*. New York: Guilford.

Peck, Jamie and Theodore, Nik (2000) ""Contingent Chicago: restructuring the spaces of temporary labor, *International Journal of Urban and Regional Research*, 24: 145–162.

Petit, Kathryn and Kingsley, G. Thomas (2003) "Concentrated poverty: a change in course." Technical report, Urban Institute, May, 180 pgs.

Philadelphia citypaper.net (2004) "Gang mentality," website http://citypaper.net/articles/2004-12-16/cover.shtml

Philadelphia Inquirer (2005) "Gangs have long been a public focus," 10 February, p. 11.

Pierce, Neil (2003) "State and local finances," technical report, Washington Post Writers Group.

Pile, Steve, Brook, Christopher and Mooney, Gerry (1999) *Unruly Cities?* London: Routledge.

Plann, S. (2004) Discussion with Planner, City of Cleveland, 9 April.

Pluralism Project (2004) "Religious diversity news," technical document, Committee on the Study of Religion, Harvard University.

Porter, Michael (1997) "New strategies for inner-city economic development," *Economic Development Quarterly*, 11: 11–27.

Professional Experience (2003) Technical document, website www.kig.harvard.edu/virtual/booktour/Old_Tour_Files/goldsmithbio.htm

Pulido, Laura (2000) "Re-thinking environmental racism: white privilege and urban development in Southern California," *Annals of the Association of American Geographers*, 90: 12–40.

Rafsky, William (1978) Discussion with Professor and Ex-Head of Urban Renewal Authority for City of Philadelphia, Temple University, 21 February.

Reagan, Ronald (1982) State of the Union Speech, January 19, Washington D.C. Available at Reagan Library.

Reed, Fred (2002) "Hopeless but not critical: the problem of race in America," *TheirOwnSelf*, 4 pages.

Reid, P. (2004) Discussion with Planner, City of Chicago, 19 August.

Reinheimer, Ira (2005) "Whose turn is it?" *Philadelphia Inquirer*, 3 May, p. 14.

Rendell, Ed (2003) Comments in Chambernet Newsletter, January–March. Website www.cbicc.org/**newsletters**/julyaugsept03.pdf

Rentgen, Mark (2004) Discussion with Planner, City of Cleveland, 12 August.

Riverwest Currents (2003) "Norquist's legacy and the new urbanism," 2, 10, October.

Roe, D. (2004) Discussion with Planner, Department of Planning and Development, City of Chicago, 6 August.

Rose, Harold (1971) *The Negro Ghetto*. New York: McGraw-Hill.

Rotberg, Robert I. and Myers, Sondra (1993) The future of Pennsylvania's small cities, research report, Lafayette College, available from authors.

Routledge, Paul (1993) *Terrains of Resistance: Nonviolent Social Movements and the Contestation of Place in India*. Westport, Conn: Praeger.

Rustack, P. (2002) Discussion with local activist, City of Cleveland, 14 November.

Salon, J. (1999) "A question of faith," 30 August, p. 18.

Schlosser, Eric (1998) "The prison-industrial complex," *Atlantic*, December 14–23: 12–19.

Schneider, Dorothea (1988) "I know all about Emma Lazarus: nationalism and its contradictions in congressional rhetoric of immigration restrictions," *Cultural Anthropology*, 13: 82–99.

Scrimger, Kay and Everett, Carol (1999) "Integral role of mayors in the international economy," *U.S. Mayor Newspaper*, 66, October, 17, 66–67.

Seccombe, Karen (2000) "Families in poverty in the 1990s: causes, consequences, and lessons learned," *Journal of Marriage and Family*, 62: 1094–2205.

Shah, Anup (2005) "High military expenditure in some places," website www.globalissues.org/Geopolitics/Armstrade/Spreading.asp

Sherman, Paul (2002) "Bush Welfare Plan: a Draconian attack on the poor," *World Socialist Web Site*, as/boomframe.jsp

Shulman, Beth (2003) The Betrayal of Work: How Low Wage Jobs Fail 30 Million Americans (New York: New).

Skrabec, J. (2004) Discussion with Worker, Department of Community Development, City of Cleveland, 5 June.

Smith, Neil (1984) *Uneven Development*. Oxford: Blackwell.

Smith, Neil (1996) *Gentrification and the Urban Frontier*. Oxford: Blackwell.

Smith, Neil (1999) "Which new urbanism: New York in the revanchist 1990s." In R. A. Beauregard and S. Body-Gendrot (eds), *The Urban Moment: Cosmopolitan Essays on the Late Twentieth Century City*. Newbury Park: Sage, pp. 185–208.

Smith, Neil (2002) "New globalism, new urbanism: gentrification as global urban strategy," *Antipode*, 34: 212–231.

Smith, Van and Segal, Fred (2001) "Can Mayor O'Malley save ailing Baltimore," *City Journal*, Winter, 2 pgs.

Smyth, C. (2004) Discussion with Planner, City of Cleveland, 19 July.

Solomon, Arthur (1974) *Housing the Urban Poor*. Cambridge: MIT.

Sorbin, Beth (2000) "NOW activists featured in health and poverty videos," website www.now.org/net/spring – 2000/outriders.html

Soros, George (1998) *The Crisis of Global Capitalism: Open Society Endangered*. New York: Basic.

Sowell, Thomas (1984) *Civil Rights: Rhetoric or Reality?* New York: William Morrow.

Sowell, Thomas (2002) "The I.Q. exemption," *Jewish World Review*, 4 March, p. 13.

Speare, Allen (1969) *Black Chicago*. Chicago: University of Chicago.

Spivak, M. (2004) Discussion with health care official, Rush Presbytarian Hospital, City of Chicago, 17 July.

Stallybrass, Peter and White, Allan (1986) *The Politics and Poetics of Transgression*. Ithaca: Cornell University.

State of Ohio Council for Economic Opportunity (2003) "Summary of current economic trend measures: Cuyahoga County, Cleveland PMSA," technical report, State of Ohio, Columbus.

Steinfels, Peter (1979) *The Neoconservatives*. New York: Simon and Schuster.

Stettner, Andrew and Allegretto, Sylvia (2005) "The Rising Stakes of Jobloss: Stubborn Long-term Joblessness Amid Falling Unemployment Rates," Briefing Paper #162, Economic Policy Institute, available from Policy Institute.

Still, Bayard (1974) *Urban America*. New York: Little, Brown, and Company.

Street, Paul (2003) "Deep poverty, deep deception: facts that matter beneath the imperial helicopters," *Dissident Voice*, 18 June.

Sugrue, Thomas J. (1993) *The Origins of the Urban Crisis: Race and Inequality in Postwar Detroit*. Princeton: Princeton University.

Swyngedouw, Eric (1992) "The mammon quest: glocalization, interspatial competition and the monetary order: the construction of new States." In M. Dunford and G. Kafkales (eds), *Cities and Regions In the New Europe*. London: Bellhaven, pp. 39–62.

Swyngedouw, Erik, Moulaert, Frank and Rodriguez, Arantza (2002) "Neoliberal urbanization in Europe: large-scale urban development projects and the new urban policy." In Neil Brenner and Nik Theodore (eds), *Spaces of Neoliberalism*. Oxford: Blackwell, pp. 195–229.

Tabb, William (1974) *The Black Ghetto*. New York: Basic.

Taylor, Humphrey (2000) "The public tends to blame the poor, the unemployed, and those on welfare for their problems," Harris Poll #24, 3 May.

Teaford, John (1990) *The Rough Road to Renaissance: Urban Revitalization in America 1940–1985*. Baltimore: Johns Hopkins.

Terkel, Studs (1970) *Hard Times: An Oral History of the Depression*. New York: New Press.

Thomas, S. (2004) Discussion with housing and neighborhood activist, Chicago, 7 June.

Thrift, Nigel (1995) "A hyperactive world." In R. J. Johnston, P. J. Taylor and M. J. Watts (eds), *Geographies of Global Change*. Oxford: Blackwell, pp. 18–35.

Tompkins, J. (2004) "School suspension up sharply – blacks suspended more often than whites, website www.poynter.org/dg.lts/id.2/aid.1682/column.html

Tonry, Michael and Morris, Norvel (1984) *Between Prison and Probation: Intermediate Punishments in A Rational Sentencing System*. Oxford: Oxford University.
Travel Business Roundtable (2005) "TBR Board of Directors," technical document from the United States Conference of Mayors, available from organization.
Triozzi, Robert (2005) Website "Triozzi For Mayor," www.http blog01.kintera.com/Triozzi/
Udesky, Laurie (1987) "Workfare: it isn't fair and it doesn't work," *Muckraker*, 1 December, website www.muckraker.org/inv_by_topic.php?topic_id=11 &&show_all=1'
U.S. Census Bureau (1940) Population and Economic Statistics, Department of Commerce, Washington D.C.
U.S. Census Bureau (1950) Population and Economic Statistics, Department of Commerce, Washington D.C.
U.S. Census Bureau (1960) Population and Economic Statistics, Department of Commerce, Washington D.C.
U.S. Census Bureau (1990) Population and Economic Statistics, Department of Commerce, Washington D.C.
U.S. Census Bureau (2000) Population and Economic Statistics, Department of Commerce, Washington D.C.
U.S. Census Bureau (2002) "Federal aid to States for Fiscal Year 2002," technical report, May, available from Bureau.
U.S. Conference of Mayors (2004) "Mayors launch new center for faith-based and community initiatives," press release, 21 January, Washington D.C.
U.S. Department of Health and Human Services (2000) "U.S. Welfare Caseloads Information," January, technical document, available from Department.
U.S. Department of Housing and Urban Development (2003) "HUD Memorandum" technical document, 12 December, Washington D.C., available from Department.
U.S. News and World Report (1975) "Any way you look at it: the worst slump since the 1930s," March 31, 26–31.
Vergara, Camilo (1994) *The New American Ghetto*. New Brunswick: Rutgers University.
Wacquant, Loic (2002) "Deadly Symbiosis," *Boston Review*, April/May, 12 pgs.
Wacquant, Loic (2002a) "From slavery to mass incarceration: rethinking the race question in the U.S.," *New Left Review*, 13: 41–104.
Wagner-Pacifici, Robin (1994) *Discourse and Destruction*. Chicago: University of Chicago.
Walkowitz, Judith (1993) *City of Dreadful Delight*. Chicago: University of Chicago.
Waller, Margy (2002) "New York program wrong for U.S.," *Los Angeles Times*, 21 April, p. 19.
Waller, Margy (2003) "Block-grant mania: a way to cut aid to the working poor?" *Philadelphia Daily News*, 28 July, p. 14.
Walters, B. (2004) Discussion with Planner, Department of Planning and Development, City of Chicago, 28 June.
Ward, Kevin (2000) "A critique in search of a corpus: re-visiting governance and re-interpreting urban politics," *Transactions of the Institute of British Geographers*, 25: 169–185.
Warf, Barney and Holly, Brian (1997) "The rise and fall and rise of Cleveland," *Annals of the American Academy of Political and Social Science*, 51: 208–221.
Washington Outlook (2003) "Can Bush finally finish off the Great Society?" *BusinessWeekOnline*, 2 pgs.

Watkins, T. H. (1994) *The Great Depression*. New York: American Heritage.

Weber, Rachel (2002) "Extracting value from the city: neoliberalism and urban redevelopment." In N. Brenner and N. Theodore (eds), *Spaces of Neoliberalism*. Oxford: Blackwell, pp. 172–194.

Weicher, John (1970) *Urban Renewal*. Washington: Urban Land Institute.

Weiler, Michael and Pierce, W. Barnett (1992) *Reagan & Public Discourse in America*. Tuscaloosa: University of Alabama.

Weinberg, Bill (2005) "Rudolph Giuliani's quality of life police state," *The Shadow Magazine*, 44, June: 6–9.

Weinberg, Daniel (2003) "Funding cuts threaten city year," *Philadelphia Daily News*, 29 July, p. 2.

White House Office of Faith-Based and Community Initiatives (2002) Text from website www.whitehouse.gov/infocus/faith-based

Wideman, John Edger (1995) "Doing time, marking race," *The Nation*, 30 October.

Wilks, M. (2005) Discussion with Planner, City of St. Louis, 15 April.

Williams, Brian (2002) "New bi-partison Welfare Bill tightens Workfare assault begun by Clinton," *Militant*, 66, 10 June: 1–3.

Williams, Kenny (2004) Discussion with housing and community activist, City of Cleveland, 5 August.

Willis, G. (2004) Discussion with activist, South Bronx, New York City, 10 July.

Wills, C. (2004) Discussion with homeless activist, 25 October.

Wilson, Bobby (2000) *Race and Place in Birmingham: the Civil Rights and Neighborhood Movements*. Rowman & Littlefield: Lanham, MD.

Wilson, David (1983) *The Spatial Character of Housing Abandonment: Philadelphia's Allegheny West*," unpublished M.A. thesis, Temple University.

Wilson, David (1993) "Everyday life, spatiality, and inner city disinvestment in a U.S. city," *International Journal of Urban and Regional Research*, 17: 578–594.

Wilson, David (2004) "Making Historic Preservations in Chicago: Space, Discourse, and Neoliberation," *Space and Polity*, 8, 43–59.

Wilson, David (2005) *Inventing Black-On-Black Violence: Discourse, Space, Representation*. Syracuse: Syracuse University.

Wilson, David (2006) "The New Segregation in U.S. and U.K. Cities." In J. R. Short, P. Hubbard, and T. Hall (eds), The Compendium of Urban Studies (Newbury Park: Sage) in press.

Wilson, David and Grammenos, Dennis (2000) "Spatiality and urban redevelopment movements," *Urban Geography*, 21: 361–371.

Wilson, David and Wouters, Jared (2004) "Spatiality and growth discourse: the restructuring of America's Rust Belt Cities," *Journal of Urban Affairs*, 25: 123–139.

Wilson, James Q. (1966) *Urban Renewal: the Record and the Controversy*. Cambridge, Mass: MIT.

Wilson, William Julius (1987) *The Truly Disadvantaged*. Chicago: University of Chicago.

Wilson, William Julius (1996) *When Work Disappears*. Chicago: University of Chicago.

Wimsatt, William (1998) "The Fear Economy," *Adbusters Magazine*, #21, Spring 10–12.

Winters, Stanley (1979) *From Riot to Recovery: Newark After Ten Years*. Boston: Rowman & Littlefield.

Wristen, Walter (1994) "Advice for the new Mayor," *City*, Winter: 11–14.

Zukin, Sharon (1995) *The Cultures of Cities*. Oxford: Blackwell.

Zunz, Olivier (1982) *The Changing Face of Inequality*. Chicago: University of Chicago.

Index

Page numbers in *Italics* represent Tables